# Introduction
# to
# PLCs

# Introduction
# to
# PLCs

## SECOND EDITION

## Jay F. Hooper

CAROLINA ACADEMIC PRESS
Durham, North Carolina

ISBN 1-59460-331-6
LCCN 2006932360



CAROLINA ACADEMIC PRESS
700 Kent Street
Durham, North Carolina 27701
Telephone (919) 489-7486
Fax (919) 493-5668
www.cap-press.com



Printed in the United States of America

# Contents

# Preface

This book is intended for the people working on and doing the day to day troubleshooting on the factory floor, and those interested in learning how PLCs work. The material in this book is presented in a format so that someone with no prior knowledge of PLCs (just some motor control, electronic, or computer exposure) can be successful in developing a good understanding of the issues and concepts involved in the workings of PLCs.

Although the examples use AB type numbering, the text is oriented to a middle of the road approach to understanding PLCs, regardless of the specific type of PLC that you use.

This book grew out of a course designed to get people from a wide variety of educational backgrounds (machine fixers and electricians through engineers) up to speed on PLCs. Almost any large or expensive piece of equipment these days comes into the workplace with a PLC attached. Close to 100% of ads for maintenance mechanics, control electricians, and manufacturing engineers require PLC knowledge.

While some generalizations have been made in the text, they are for the purpose of enhancing the overall understanding of the PLC material presented.

Jay F. Hooper
Greensboro, NC

# Foreword

This introductory book on PLCs is oriented to the line mechanic and the electrician working on the factory floor. It is directed to their world view (series and parallel). However, I have included logic circuit equivalents from the computer world and the electronic world (AND, OR, NOT, NOR NAND, etc.) in an Appendix for Chapter 4.

The book does not go into the design of PLCs or the design of PLC systems as this would be beyond the scope of an introductory book.

The material presented would be most useful as a text for community college courses (both curriculum and continuing education) for electrical programs, industrial systems programs, or industrial maintenance programs. It would also be highly useful as a lab manual for four year college or university electrical, electronic, or systems courses.

Trouble shooting of PLC programs and problems is usually accomplished in labs. Software glitches are usually handled on a vendor specific basis and hardware glitches (using volt-ohm meters, etc.) is usually covered in a prerequisite course such as motor controls or during on the job training.

The various aspects of this approach have been thoroughly tested over a wide range of audiences over the past dozen years.

Jay F. Hooper
Greensboro, NC

# Introduction to PLCs

# Overview 1

What is a programmable logic controller (PLC)? It is basically a regular computer (PC) that's been reduced to a micro-sized package. This package has been industrialized as to its temperature range and electrical noise characteristics.

The distinguishing thing about the PLC is that it doesn't use individual wires, like a legacy PC parallel port or a mouse port, which run on 5 to 12 volts DC to manipulate a device. The "wires" on a PLC are instead hooked to individual terminals that work on, say, 120 volts AC. Ouch! Needless to say you need to be careful where you put your fingers around the terminals of a PLC. It is not unusual to have up to 220 VAC devices hooked to an input or to an output terminal.

PLCs are used extensively to control devices. Equipment includes the standard conveyor/packaging industrial-type line down through medium-sized air conditioning units. PLCs can even be found potted on circuit boards of high end washers and dryers for the home.

The two most requested hands-on courses in the controls area are PLCs and motor control. "Everybody and their uncle" wants to know how to work on PLCs. One reason is that almost every new machine that comes into a factory has a PLC on it. Another is that almost every electrician and industrial maintenance job advertised for industry has knowledge of PLCs as a job requirement.

## An Origin Story

So how did this control solution called a PLC come into such widespread use? Well, let me tell you an origin story to give you a sense of what happened.

At one point in the history of the car industry there were a lot of sheet metal changes every model year. This necessitated frequent changes in the configurations of the machines used in automobile manufacturing plants. The limit switches and sensors were hooked to banks and banks of control relays. These, in turn, had to be hand-wired every model year.

One year someone at a car company realized that there was "a whole lot of switchin' going on." That person thought that maybe the company could use a mini-computer to manage the interfaces from the switches and the sensors to all of the solenoids and contactor coils. That way, the company would only need to wire up the sensors and the coils one time, and just change the logic program in the mini-computer each model year.

So, the company requested designs from various mini-computer manufacturers, who developed rudimentary PLCs and installed them in the factory. Well, after a period of time the company met with the mini-computer folks and said, "We have some good news and some bad news. The good news is that the units worked OK in our factory applications. The bad news is that we can't use any of them."

"What?" "You've got to be kidding." "What's the problem?"

It turned out that all of the units were using a high-level computer language such as FORTRAN or a low-level language such as Assembler to run the mini-computers. The problem was that in order for factory floor workers or troubleshooters to make a change or a modification in the program, they would have to know the programming language or they were stuck.

Someone in the company mused, "Well, you know, all of our electricians and control people know ladder logic. Now if the units could be programmed in ladder logic...."

"The rest is history," as they say.

That brings us to the state of the PLC in the present day.

## The Four Parts of a PLC

Almost every book on the subject will tell you incorrectly that the PLC consists of three parts, the input rack, the output rack, and the CPU. Of course, you want to know how PLCs really work, don't you?

The PLC consists of four parts

- the input rack,
- the output rack,
- the internal rack, and
- and the CPU,

that are related like this:

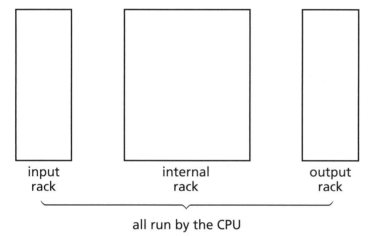

input rack          internal rack          output rack

all run by the CPU

**Figure 1.1. The basic components of a PLC.**

The input rack has sensors hooked to it, such as pushbuttons (PBs), limit switches (LSs), photo eyes (PEs), switches (Ss), etc.

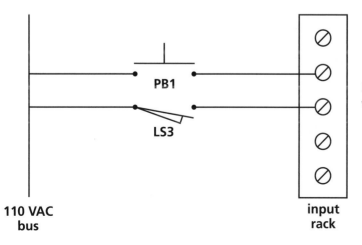

**PB1**

**LS3**

**110 VAC bus**

input rack

**Figure 1.2. Sensors wired to an input rack**

The output rack has coils or loads hooked to it, such as solenoids (SOLs), contactors (CONs), pilot lights (PLs), motor starters (MSs), etc.

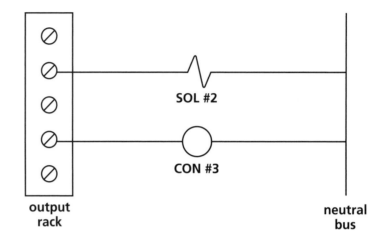

**Figure 1.3. An output rack with two loads.**

The internal rack consists of software, and has items such as control relays (CRs), timers (TONs and TOFs), counters (CTUs and CTDs), sequencers (SQOs), etc. hooked to it.

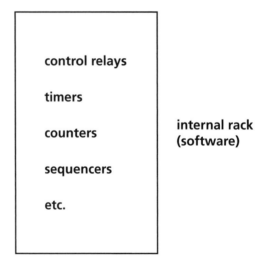

**Figure 1.4. The internal rack.**

The CPU runs the whole thing.

## PLC Symbols

In order to do our first exercise we need to know about the PLC symbols and how they came into being. At the time that PLCs were being developed, a choice had to be made about the symbols that represent the various elements of PLCs. The PLC is an industrial device put together by committee. Electricians, electronic techs, computer types, and engineers all wanted to have the symbols and the numbering systems from their world on the PLC.

At that time, a computer graphics terminal and its specialized keyboard were way too expensive to use on a PLC control unit. The choice was made to use CRTs and the ASCII keyboard. Now you can stare all day at your ASCII keyboard, but you would be hard pressed to find a three input OR gate symbol on it. So the PLC uses ASCII characters to create symbols and terminal numbers.

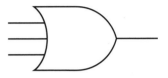

**Figure 1.5. Symbol for a three input OR gate.**

The result of choosing the ASCII keyboard as the "symbol maker" for the electrical and electronic worlds is that it all is a bit confusing. When you first start to learn how to program PLCs, having to create the symbols makes the programming seem a little crazy. The great thing about ASCII, though, is that from your desktop through a dial-up modem you can control devices anywhere in the world. A company headquartered in Raleigh, North Carolina can control building systems (AC, hot water, lights, etc.) in Boston or Los Angeles.

As the cost of technology has come down, graphic displays are now affordable enough that you will encounter them on the job. This book uses both the original ASCII symbols and drawn graphic images to represent PLC symbols.

## PLC Terminal Numbers

The bit number system for PLC terminals (you know, a bit, a nibble, a byte, and then a word, lunch anyone?) usually runs something like this:

WHAT I AM / LOCATION / TERMINAL #.

An example of an output rack terminal number is

$$0:2/8$$

where O is for output rack, :2 is for the location (slot #), and /8 is for the terminal #.

An example of an input rack terminal number is

$$I:1/3$$

where I is for input rack, :1 is for the location (slot #), and /3 is for the terminal #.

An example of an internal rack terminal number is

$$B3/4$$

where B3 is for internal control relay output or input, and /4 is for the terminal #.

Now remember, not all PLC inputs start with I. Not all PLC outputs start with O. Not all internal control relays start with B3. Each manufacturer has its own way of doing things.

## PLC Logic Symbols

The first symbol that we will look at is the one for examine on (EON):

$$-]\ [-$$

This symbol can mean two different things.

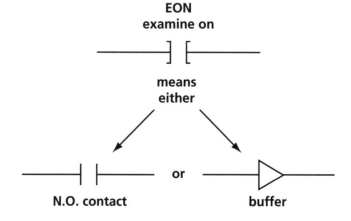

**Figure 1.6. Examine on symbols**

The meaning depends on the rack (input, output, or internal) that the terminal # is associated with.

If an examine on has an output rack # or an internal rack #, the EON mimics (is the PLC equivalent of) a normally open (NO) contact. We'll be talking more about normally open contacts later.

```
  O:2/1              B3/1
  -] [-              -] [-
   ms1                cr3
```

**NO contacts**

These are normally open contacts. If they are closed, they pass along a signal from the left of the contact to the symbol to the right of the contact; if open, they pass along no signal to the symbol to the right of the contact.

If an examine on has an input rack #, the EON mimics a buffer or a YES gate. We'll be talking more about buffers and YES gates later.

```
  I:1/5
  -] [-
   pb2
```

**buffer or YES gate**

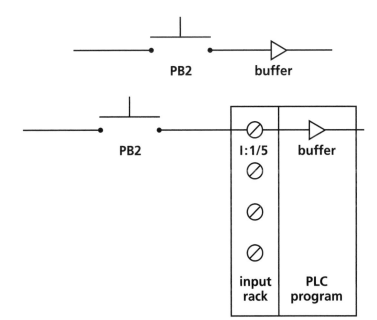

**Figure 1.7.**

The second symbol that we will look at is the one for examine off (EOF):

$$-]\ /\ [-$$

This symbol can mean two different things.

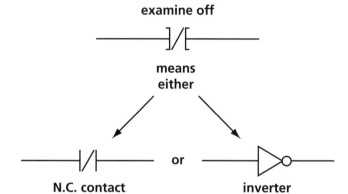

**Figure 1.8. Examine off symbols.**

The meaning depends on the rack (input, output, or internal) that the terminal # is associated with. Sound familiar?

If an examine off has an output rack # or an internal rack #, the EOF mimics a normally closed (NC) contact. We'll be talking more about normally closed contacts later.

```
   0:2/2              B3/2
  -] / [-            -] / [-
   ms1                cr3
```

**NC contacts**

If an examine off has an input rack #, the EOF mimics an inverter or a NOT gate. We'll be talking more about inverters and NOT gates later.

```
   I:1/6
  -] / [-
   pb3
```

**inverter or NOT gate**

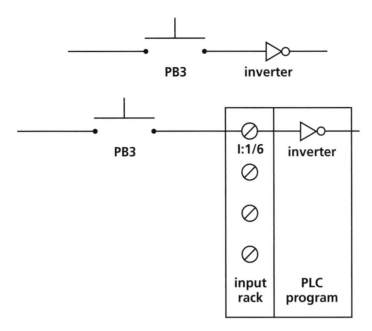

Figure 1.9.

The main thing is to realize is that EON and EOF can mean two different things. The meaning depends on which rack the symbol is associated with. These meanings are from the electrical, electronic, and computer worlds. Ah, yes, the ASCII world at work.

The previous two symbols were inputs. We also need an output in order to do our exercise for this chapter. The third symbol we will look at is output energize (OUT):

$$-\langle\ \rangle-$$

This symbol means just one thing, energize the output. You will find it both on the internal rack and on the output rack.

```
B3/4            O:2/4
-( )-           -( )-
 cr4             ms2
```

**output energize (OUT)**

Please turn to and do Exercise #1 before continuing with the next chapter.

# Hardware 2

This chapter will introduce you to the hardware associated with PLCs.

## Input Rack

First of all, let me show you how the input rack works. Figure 2.1 shows a number of sensors hooked up to the input rack terminals:

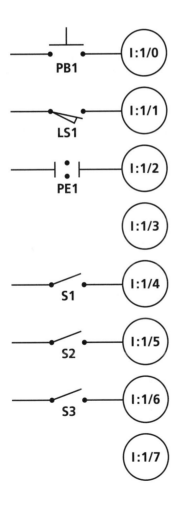

**Figure 2.1.**

All of these sensors are hooked to a 110 volt AC bus, as shown in Figure 2.2.

**Figure 2.2.**

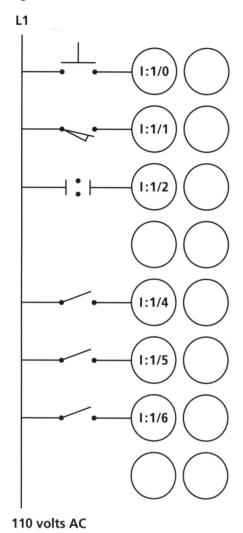

The bus is connected to the sensors, such as pushbuttons, photo eyes, limit switches, etc. The 110 volts goes through the sensors and to the terminals of the input rack.

On the input side of a typical PLC arrangement you would have sensors that run at 110 volts AC, instead of 12 VAC or 480 VAC. Of course, the CPU and associated logic circuitry in the PLC runs on 5 volts DC. You have to go through a voltage level shift in order for the computer to read the inputs correctly. After all, if you put 110 volts AC on a 5 volt DC system, you will send sparks flying everywhere and short out the PLC computer.

So you go through an interface that shifts the 110 volt AC signal to a 5 volt DC signal. Typically this shift is done by running the signal through a couple of dropping resistors, a bridge rectifier, and a zener diode, and then to one side of an opto isolator. Inside the opto isolator, the 5 volt DC computer of the PLC basically looks across a gap and is electrically isolated from the 110 volt AC logic levels of the input terminals. In summary, the AC sensor signal is sent to a PLC input terminal and then toggles the DC signal to the computer. It looks something like Figure 2.3 (below).

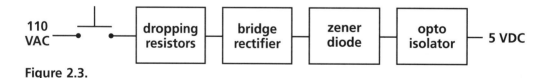

**Figure 2.3.**

## Output Rack

On the output side of the PLC you typically have a 110 volt AC neutral bus, similar to the output terminal diagram in shown in Figure 2.4.

Here you would have solenoids, lights, horns, contactor coils, and that sort of thing. The 110 volts AC goes through the individual terminals on the output card, through your loads, and then to the neutral. What happens is that the computer puts a signal on the gate side of a solid state device or on the relay coil of a small control relay on the output card. This toggles either the solid state device or the relay to ON, and puts 110 volts AC on the output terminal. The 110 VAC then goes through the load to the neutral and turns the output load on.

### *Different Voltage Levels*

On the input card you can have all sorts of different voltage levels occurring. Typically in a manufacturing plant you have 110 volt AC inputs, but you could have 220 volt AC inputs if you had a European machine, or even DC inputs if you had a system oriented towards electronic devices. Whatever input arrangement you have, the voltage must be reduced and rectified to toggle signals on the 5 VDC PLC computer.

On the output rack you have a similar arrangement. You might have 110 volt AC output terminals to turn things on and off. Of course you can have other types of output cards. In an air conditioning system, you could have an input or an output card that ran on 24 volts AC. Air conditioning systems use 24 volts AC from the thermostat to turn the contactors on and off. This in turn cycles the compressor on and off. Your air conditioner's compressor in your home usually runs on 220 volts AC single phase. You could have a PLC output card that outputs 24 volts AC for the compressor contactor coil if you have a sophisticated high-end air conditioning system.

**Figure 2.4.**

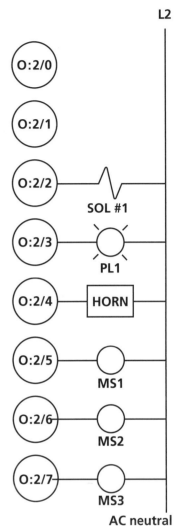

### Current Drains

Another consideration is that there is only so much current you can put through the components on the output card of a PLC. Generally speaking, you cannot hook motors directly to output cards. You might have only 350 milliamps (maximum) of current available on a particular output card, so you have to use a contactor between the output card and the motor. The size of the contactor that you can turn on and off is limited as well. You might be able to turn on and off a contactor that in turn switches a 5 to 10 horsepower motor on and off. However, if you have a contactor much bigger than that, your output card is not going to be able to do that directly. There might not be enough current available to drive the contactor coil directly. Sometimes you may need to have the output card drive a control relay or a power relay that in turn drives the contactor.

## Terminal Diagrams

The figures earlier in the book show input and output diagrams. These are called terminal diagrams, and they show whether a sensor is hooked up normally closed or normally open, etc.

It is important to realize that what appears on your screen in the PLC is not the terminal diagram but the PLC logic, which includes, for example, examine ons (EON) and examine offs (EOF). The logic looks similar to the example in Figure 2.5 (on the next page).

When you look at this figure, you will notice the output and input terminal numbers. It shows a pushbutton, a limit switch, and other types of input devices. The diagram shows motor starter coils and solenoids. However, none of this information tells you whether your sensor is hooked up normally open (NO) or normally closed (NC) to the input terminals. That information is not there. You have to use the information on the terminal diagram to determine the NO or NC status of a sensor.

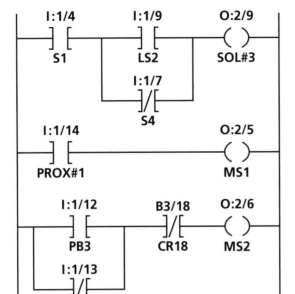

**Figure 2.5.**

You are most likely to find terminal diagrams in the packet in a pocket of the machine you are working on, or in the engineering office in the plant. I don't know what it's like at the plant where you work, but it has been my experience that sometimes when you are working on second or third shift, the terminal diagrams are nowhere to be found. If you do find one, it might look like someone ran over it with a fork lift and left tire marks all over it. Maybe one corner is burnt. That sort of thing occurs quite frequently.

Terminal diagrams are supposed to be in the pocket of the machine, but.... If you need some information about the status of a sensor hooked in the rest position, it is not going to be on the CRT screen. NO and NC information is going to be on the terminal diagram.

A typical input output terminal diagram looks like the one in Figure 2.6 on the next page.

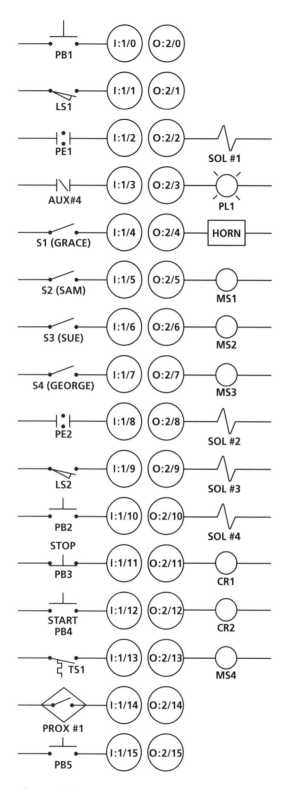

**Figure 2.6.**

## Seven General Rules

Another thing that is going to help you learn how PLCs operate is this: There are about seven general rules you need to follow to be the most successful in how you understand and work with PLCs.

In this chapter, we will talk about general rule #1 and general rule #2.

### General Rule #1

What I call the "Ivory Snow Rule": **99 and 44/100% of the time your inputs need to be hooked to buffers (examine ons) on the input rack.**

Now this may seem like a simple thing, but if you wire up a machine, or the manufacturer has sent you a machine, the sensors will come in two varieties, normally closed (NC) and normally open (NO). If you have a normally closed sensor and you run it to a buffer on the input rack, your computer is going to read it as a one. If you have a normally open sensor and you run it to a buffer on the input rack, the computer is going to read it as a zero. This is what you normally want to occur. If everything is wired properly, practically all of your sensors on the input rack are going to run to examine ons that are used as buffers.

Of course, you have examine offs available, in case you miss something or wired something up wrong. If you want to swap something around, you can do it in software much easier than going back out to the machine and rewiring it.

## General Rule #2

**You cannot put emergency stop buttons (E-stops) in the PLC. You must hard wire them.**

This concerns output cards and safety. The National Electrical Code (NEC) requires motors to have a separate way to disconnect from the power source. Here is what that means for your PLC: you are not allowed to put the emergency stop button in the PLC. You may put a start and stop button for a motor in the PLC. That's not a problem. But the emergency stop button needs to be hard wired. You can go to Fortune 100 companies around your area today, and I guarantee you that half of the plants you visit will have emergency stops in the PLC.

The problem with putting an emergency stop in the PLC is that the PLCs have racks of solid state devices that toggle on and off, and they in turn toggle loads on and off. If you have an emergency stop hooked into the PLC, and have the motor starter coil hooked to a triac, you may have a very bad situation. The problem you could run into is that when someone tries to turn off the machine, the solid state device might fail in the ON position. Someone hits the emergency stop. If the emergency stop is in the PLC and the triac fails in the ON position on the output card that controls that contactor coil, then the machine is not going to stop. Someone could get hurt or killed.

So what you need to do as far as safety is concerned is to put the stop and start button in the PLC, but you need to hard wire the emergency stop. On the output card you would have a triac with main terminal one (MT1) and terminal two (MT2), sort of like this:

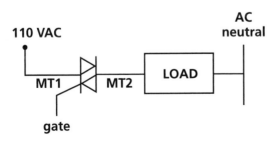

**Figure 2.7.**

The gate gets toggled by the 5 volt DC associated circuitry from the PLC computer. When the triac gets this signal, 110 volts from the main terminal goes out into the output coil. To get the 110 volts AC, a wire runs from the 110 volt line to the output card. The 110 VAC goes through the main terminals, and then out to activate your contactor coil. For the emergency stop, what people typically do is put a interposing relay in the 110 volt line, like this:

**Figure 2.8.**

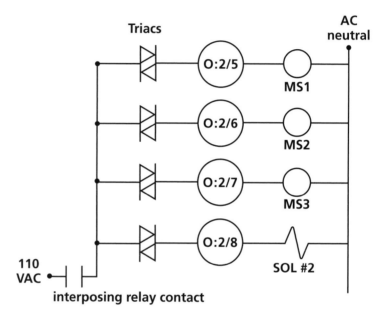

Then, when someone pushes the emergency stop button, it is going to drop out this relay and that is going to kill the power to all the output terminals on the output card. Consequently, any motors that are connected here will drop out and stop, because you are killing power to the motor contactor coils.

This is an important safety issue. The second rule should always be followed: *when you have an emergency stop on your machine, never put the emergency stop in the PLC.* Always hard wire the emergency stop, whether you are using an interposing relay, which we will discuss later, or some other method. You must have a separate emergency disconnect to stop a machine to be in compliance with the NEC.

Please turn to and do Exercise #2 before continuing with the next chapter.

# Programming Basics **3**

PLC programming basics begin with making a clear distinction between inputs and the input rack, and between outputs and the output rack. It is important to remember that inputs are different from what is on the input rack. Typically, a beginning level person thinks they are the same, but they are not. The same situation applies to the outputs and the output rack. What an output is and what is on the output rack are two different things.

## Inputs and Outputs

Most sensors and everything on the input rack are, of course, inputs. But you might also have a situation where the internal rack has some control relays. Control relays have control relay coils, which are considered outputs. Control relays can also have a normally open contact with the same address, which is used as an input. Notice B3/1 in Figure 3.1:

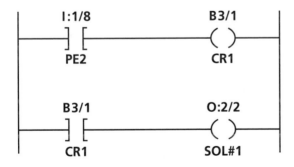

**Figure 3.1.**

The general rule to keep all of this straight is that the outputs tend to be in the very last column on the extreme right on your CRT screen. Everything in front of this (in the middle and on the left part of your screen) tends to be inputs.

Another example is an input that has an output rack number. In Figure 3.2 you have a normally open contact, MS2 at O:2/6, that mimics a motor starter coil auxiliary contact. It is on the left side of your screen and it acts as an input to the solenoid coil, SOL #2 at O:2/8.

**Figure 3.2.**

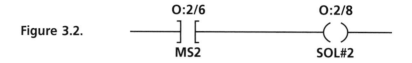

O:2/6                                    O:2/8

MS2                                      SOL#2

The main thing to remember is that there is a difference on the PLC between the term 'input' and what's on the input rack, and the term 'output' and what's on the output rack. Again generally speaking, on the CRT screen everything on your extreme right hand side is an output and everything preceding it is an input.

## The IO Scan

Another issue that comes into play when you're working with PLCs is the IO scan. If you wire up a regular circuit (a hard wired circuit), basically everything works at the speed of light (instantaneously as far as you or the machinery is concerned). On the PLC it appears that everything works instantaneously, but you have to remember that the PLC actually reads the program one line at a time and also updates its files that contain information about the status of the inputs and outputs. Now of course it reads the whole program very fast. You may have a 783 line program and it may read the whole program in one-sixtieth of a second, but the fact of the matter is that the program is going to update what is happening on line number 7 before it gets down to line number 111, even through the program runs very quickly.

So the problem you run into occasionally is that in one line the computer is doing something to an item, and then in less than one-sixtieth of a second the computer is doing the opposite thing to the same item in another line. This is a glitch in your program. So the third general rule I need to tell you about when you work on your PLC is this:

### General Rule #3

Only use a particular output one time in your program. You won't put an output, such as SOL #3 at O:2/9, in line number 7 and have the output again in line number 111, as shown in this figure of a problem:

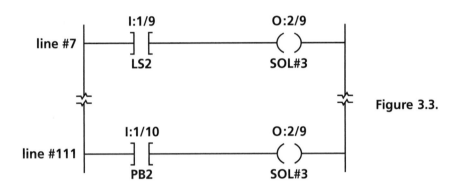

Figure 3.3.

The reason you don't want to do this is to avoid creating little glitches in your program, where the output on one line overrides the same output on another line. You can prevent this by using the output just one time. If more than one condition affects a particular output in your program, then you should put the output on one line and regulate it by using AND and/or OR conditions as illustrated in Figure 3.4:

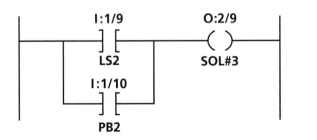

Figure 3.4.

So a good general rule to follow to keep you from having glitches in your program is: use an output just one time in your program.

## Branching

Another consideration in programming PLCs is the branching situation. For most people, the easiest way to do branching is to follow the original ladder logic that an electrician would use. In order to run the machine, the electrician would basically put things in series or parallel. But a thing that you can do while hard wiring a circuit that you can't do with PLCs is vertical runs.

Basic horizontal runs of course look like this:

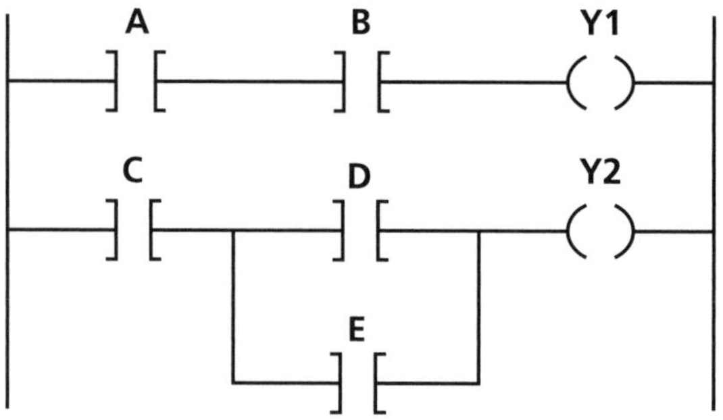

**Figure 3.5.**

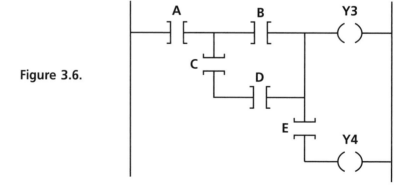

Vertical runs would look like C and E in Figure 3.6:

**Figure 3.6.**

Of course, there is no problem hard wiring vertical runs when you are actually doing point-to-point wiring. Vertical runs are not allowed, however, in a PLC program. The next general rule for PLCs is:

### General Rule #4

**Don't use any vertical runs.**

If you have a vertical rung situation similar to the one shown in the hard wired circuit in the previous figure, then on a PLC you would actually have to program it like this:

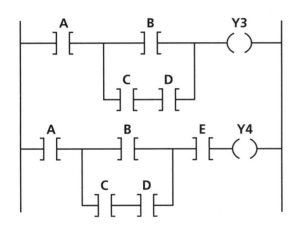

**Figure 3.7.**

## Gates

The major gates that you will encounter on PLCs are YES, NOT, AND, and OR. Actually, you don't use the gates themselves, you use the PLC equivalents of the gates.

You need to remember how to hook up buffers and inverters on the PLC. Buffers are equivalent to YES gates. Inverters are equivalent to NOT gates.

I already told you about the Ivory Snow rule: almost 100% of the time you are going to hook up sensors to buffers on the input rack. That's something you are generally going to follow even if you have a fairly sophisticated operation. Generally speaking, if you go into a plant you should see that most of the input sensors are hooked to buffers, as opposed to going into a

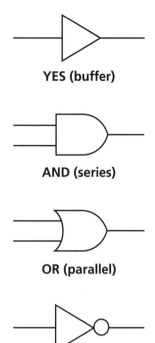

**YES (buffer)**

**AND (series)**

**OR (parallel)**

**NOT (inverter)**

**Figure 3.8. Gate symbols.**

plant and seeing that most input sensors are hooked to both buffers and inverters. Finding that most of the input sensors are hooked to buffers tells you that more thought and a little more professionalism went into the machine.

That's not to say that you can't use inverters. Inverters are available to the machine builders and troubleshooters primarily to keep you from having to go out and rewire something that was hooked up backwards.

The other two major gates that you find on PLCs are AND gates (on most PLC programs, you are going to hook things up in series to mimic an AND gate) and OR gates (on most PLC programs, you are going to hook things up in parallel to mimic an OR gate).

## Symbols on Terminal Diagrams

Another thing to consider, particularly if you are from the electronic, the computer, or the engineering worlds, is electrical symbols and what they mean on terminal diagrams. I have put a group of the most commonly used symbols and what they mean in the appendix. You should familiarize yourself with them.

So to summarize:

- Don't use any vertical runs.
- Only use an output one time in your program.
- Buffers are equivalent to YES gates.
- Inverters are equivalent to NOT gates.
- Hooking things in series is equivalent to AND gates.
- Hooking things in parallel is equivalent to OR gates.

Please turn to and do Exercise #3 before continuing with the next chapter.

# Basic Logic 4

This chapter is an introduction to logic, especially the major logic functions (AND, OR, NOT, OUT, NAND, and NOR) that you are normally going to run into while working with PLCs. We will be looking at how logic works in our everyday speech and how it works on PLCs.

The primary thing to note when you are dealing with logic functions is that most problems can have several different solutions. There is the traditional approach of using gates, such as those used for electronics, piping, and process control. You may even use Boolean algebra. However, in PLCs you generally do not use logic gates. When PLCs were originally developed, people were not using graphics terminals because those terminals were prohibitively expensive. Consequently, they just used the symbols on a regular keyboard to represent the logic they needed.

The AND "gate" and other "gates" that were developed for PLCs were set up to be familiar to electricians, because they were most likely to work on them on the factory floor. In this book, we are going to approach logic from the electrician's viewpoint, because that is what you are going to run into on the factory floor the vast majority of the time. (The computer world and the electronic world logic gate equivalents for this chapter are in Appendix B.)

## AND and OR

The AND "gate" developed from the electrical series circuit. You hook two items or more in series, as shown in Figure 4.1, to control some sort of output.

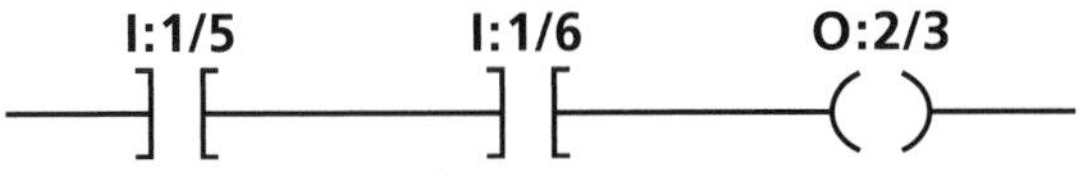

I:1/5     I:1/6     O:2/3

**Figure 4.1. Inputs connected in an AND configuration.**

If you have multiple inputs, and want any one of them trigger the same output, you would use OR logic. Two or more inputs are hooked in parallel to give the appropriate output, such as this:

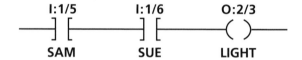

**Figure 4.2. Inputs connected in an OR configuration.**

These two logical situations are fairly straightforward. You are basically saying in the first instance (AND) that if Sam and Sue both did this particular thing then you are going to have this outcome over here, like the light coming on in this example:

**Figure 4.3. AND logic: both Sam and Sue must act to turn on the light.**

In the second example (OR), you might say something like, if Sam or George or Sue did some sort of action then you have an output, like the horn sounding in this example:

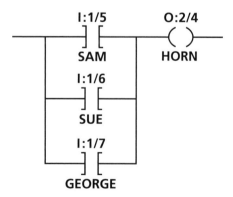

**Figure 4.4. OR logic: Either Sam, Sue or George can turn on the horn.**

It's fairly straightforward when you are hooking things in series and parallel.

## NOT

The next logical operator is a NOT gate, which is also known as "No." This is illustrated in Figure 4.5, where the input devices are sensors. If Sam is in the room and Sue is not, then the light comes on.

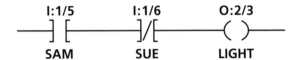

Figure 4.5. Sam but NOT Sue.

You will notice that in this configuration we are using a control scenario similar to a control relay, where you have a normally-open and normally-closed contact, or you can have an input sensor that acts as an inverter. At any rate, the input from Sam and Sue are opposite each other. Basically, the way it works is, if not Sue, but Sam, then the light comes on. If Sue is in the room, whether Sam is in the room or not, then the light is not going to come on.

A lot of times we use a normally-closed contact as a way of expressing NOT in a PLC, but you can also use inputs with an examine off as an inverter to express NOT. So you have several different options.

## OUT

The output in PLC lingo used to be called "OUT," but now most of the time people use the term "output." Remember there is a difference in what people perceive as an output and what is on the output rack. Sometimes that gets confused. Formally, an output is called OUT.

## NAND and NOR

Then of course you have the arrangement where NOT is associated with AND. This is a form of basic logic known as a NAND gate. NOT associated with OR

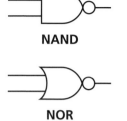

**NAND**

**NOR**

**Figure 4.6. NAND and NOR symbols.**

**Figure 4.7. NAND gate.**

**Figure 4.8. NOR gate.**

is known as a NOR gate. The logic symbols are shown in figure 4.6.

## Logic in Everyday Speech and in the PLC

In the following scenario, if you do not have Sam and you do not have Sue then you have the light on. So your program will look like this:

As you can see, in the rest position without detecting Sam and without detecting Sue, the light is going to be on.

In the next example, we have a situation where if we do not have Sam, or we do not have George, or we do not have Sue, then the horn is on.

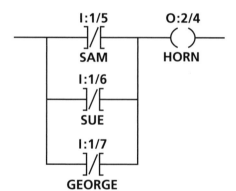

As you can see, if nobody is sensed in this particular program, the horn is going to be on. This would be called a NOR gate, and of course there are different ways you could think about this, but the easiest is to say the sentence in English and then translate it to the PLC.

In the next example, this statement: "If Sam or George, but not Sue, is detected in the room then I want the light to turn on," you could fairly easily write that in a PLC logic statement.

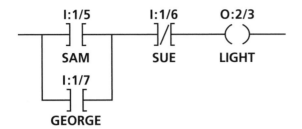

Figure 4.9.

We are saying "if Sam or George (on a PLC diagram that would be an OR situation), and no Sue." You have a situation where you need an OR, a NOT, and you also need to have an output.

You run into a similar situation on pneumatic control elements when you have the operators that are ANDed or ORed on a directional control valve. If you have a double safety of electrical and air systems, a valve might be set up so it has to have an air and an electrical signal to operate. If you pull the disconnect on either energy source, the valve is absolutely not going to work.

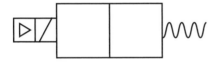

Figure 4.10. Electrical AND pneumatic pilot operated directional control valve.

You can do a logic-type approach using just ordinary speech with the proper placement of commas. The whole key is where the comma is placed.

## General Rule #5

Watch those commas.

I find that using ordinary speech is the easiest way to work for most personnel in putting logic situations into a PLC. A lot of folks are scared silly using truth tables, and do not really know how logic gates work. Certainly most people on the floor are very intimidated by Boolean algebra and find it very difficult to translate Boolean algebra into PLC logic. After all is taken into consideration, PLC logic (which basically is a series or parallel hookup with an output) is what they need for a specific item or sensor to do its job in a particular situation.

Let's try one more example using everyday speech. We have a situation where we say if Sam, and George or Sue go into the room, but not Grace, then we are going to shut the light off. Let's look at this figure to see if this circuit would do it:

**Figure 4.11.**
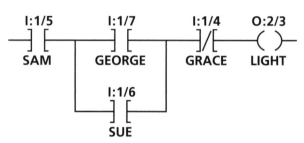

In the first part Sam, and George or Sue are going to be hooked in series, and Grace in turn is going to be hooked in series to them, and we have a situation where the light is going to be on.

The original sentence said, however, that the light needed to be off. So just take the program in Figure 4.11 and invert all the inputs like this:

**Figure 4.12.**
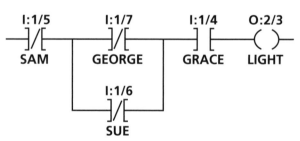

Now the program works as advertised. Another solution would be to use a control relay in a two line program like this:

**Figure 4.13.**
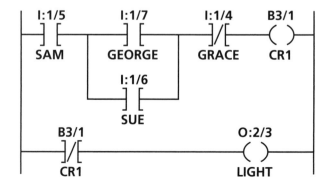

I find that the normal workers, electricians, and troubleshooters on the factory floor can get fairly sophisticated logic using everyday speech and series and parallel connections with a PLC. In most instances, everyday logic works out fairly well for sophisticated folks as well. They may know how logic gates operate, but the fact is that most PLCs have simple series and parallel hookups. It behooves everyone to try to get the action they want and still speak a common, understandable language. Anyone who has to troubleshoot or make modifications to a machine needs to be able to understand what's going on.

At this point please turn to and do Exercise #4.

# Ladder Logic 5

Most of the ladder logic on the PLC screen comes from the electrical viewpoint. That is to say, the symbols are displayed with the types of controls, sensors, and outputs that an electrician would normally run into.

## Control Relays

First, there are control relays. In the example below you will see there is a control relay contact and a control relay coil. Both are parts of B3/1 (CR1).

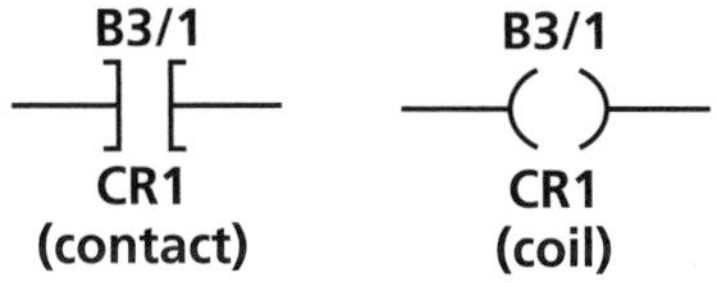

B3/1                 B3/1
─┤ ├─              ─( )─
CR1                  CR1
(contact)           (coil)

**Figure 5.1. Control relays.**

In the figure above, B3 stands for control relay, but your particular PLC might have a different number for a control relay.

Control relay components (inputs and outputs) are programmed in the PLC software on the internal rack. Basically, on PLCs the control relay bits are in the same memory space or memory map area as the sequencer bits. Usally, the bits in the top part of the memory map area for control relays, starting with control relay number one (CR1) and going to, maybe, control relay number two hundred (CR200). Typically, the bits in the bottom part of the memory map are for sequencers. (We'll talk about sequencers later.)

All manufacturers have a block of numbers dedicated to control relays. These numbers represent a part of the PLC memory map. For example, bits 701 through 999 might be where all of the internal control relays are.

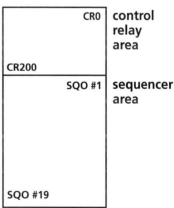

**Figure 5.2. Memory map.**

Internal control relays are used in PLC programming when you do not want to use real world physical PLC output terminals or input terminals to do your logic. You can use control relays in the internal rack to perform the operation in software as in Figure 5.3:

**Figure 5.3.**

B3/8            B3/11

CR8            CR11

Here, control relay contact B3/8 turns on or off internal relay coil, B3/11.

## Contactors and Motor Starters

As I mentioned earlier, you very rarely run into a situation where the PLC is driving a motor directly, because there is just not enough amperage available to the applicable PLC output terminal to do that. Typically, PLC output and input cards (particularly on the output cards) have a maximum current of around 350 milliamps available to drive a load.

So how does a PLC turn on a motor? A motor starter is basically a contactor with an overload relay. The PLC drives the coil on the motor starter, instead of driving the motor. The PLC can turn the motor starter coil on and off, even though the PLC couldn't turn the motor itself on and off directly. An AC motor starter coil has a horsepower rating well below the horsepower rating of the motor.

But if you have a fairly large load, like a 15 or 20 horsepower motor, you need more output than 350 ma to turn on the motor starter coil. So, there will often be a hardwired relay in the circuit used to run a larger contactor coil. The hardwired relay is called an interposing relay.

## Two-Wire (Automatic) Control Circuits

Figure 5.4 shows one of the most common ways to control a motor starter, a two-wire automatic control:

**Figure 5.4. A two-wire automatic control.**

This circuit has a sensor that needs to have some sort of input to run a fan (it could be a temperature sensor running a compressor unit, or high-end air conditioning unit on a commercial building or plant, or perhaps a position sensor). A device that automatically gives you an output with a signal is called two-wire automatic control. It is called this because there are two wires in the circuit between the sensor and the load.

The problem with two-wire automatic control circuits is safety. Say you had a thermostat hooked to a large ventilating fan in a warehouse. It is hot and the thermostat is calling for the fan to be on. If you lose power, of course the fan stops. The thermostat is still calling for the fan to be on, but because you have lost power the fan is off. Now if someone is working around the fan and the power comes back on, since the thermostat is still calling for the fan to be on, the fan will automatically start back up. This could represent a danger to people working around it. Sump pump installations have a similar circuit.

## Three-Wire (No-Voltage Protection) Control Circuits

A three-wire control circuit is safer. This circuit is also called a low voltage protection circuit, a no voltage protection circuit, a safety circuit, or a stop-start circuit. This circuit is quite common in industry. It looks similar to this:

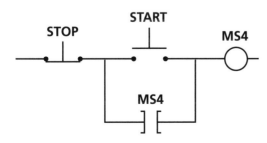

**Figure 5.5. A start-stop pushbutton station.**

In the PLC, you can program the circuit in one of two ways. The first way is to directly mimic the hard-wired electrician stop-start pushbutton latching circuit where a motor starter contactor will lock itself in. The program would look similar to this:

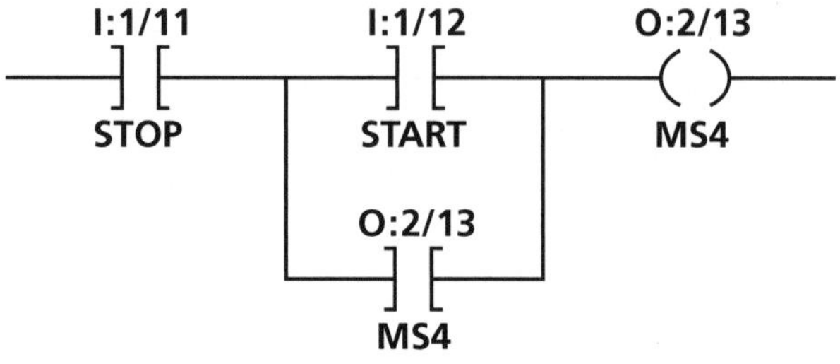

**Figure 5.6.**

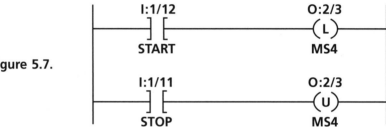

The second way is to put the circuit into the program using the latching and unlatching function on the PLC, which looks like this:

**Figure 5.7.**

Either way will work fine. There is one school of thought that likes to use one way and another school of thought that likes to use the other way, but both methods of using the PLC to control the motor starter coil (which we have a bunch of in industry) will get the job done.

## Open vs. NO and Closed vs. NC

Now let's turn to the terminology of open versus normally open (NO), and closed versus normally closed (NC). You have an interesting thing in the electrical world, if you think about it.

In just about every profession in the world, opening up a valve, a door, a gate, or opening up just about anything, permits some sort of flow through the system. But in electrical work, if you open a circuit, you stop the flow of electricity. On the other hand, in just about every system in the world, closing a door, valve, or gate impedes or stops the flow of whatever is going through the system. The exception is in electrical work. In the electrical world, closing a circuit allows the current to flow.

So in electrical work you have the situation that the English words open and close are basically turned on their head, and they mean exactly the opposite of all other professions.

Now while you have that going on, you also have a situation where a switch may be normally closed (NC) or a switch may be normally open (NO). What's that about? Well, I am going to explain further so as not to leave you confused, and also tell you how closed and open fit in.

Normally closed and normally open switches can be illustrated like this:

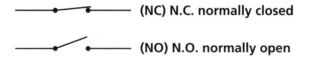

Figure 5.8. Normally closed and normally open switches.

Now, what is the difference between normally closed and normally open and just plain closed and open? The difference is what the switch is doing in the rest position (the way the switch is drawn on the print) versus what the switch is doing at any given instant.

Take, for example, the light switch in your refrigerator. The circuit would look something like this:

Figure 5.9. A normally closed, single pole switch controlling a light.

This switch is a normally closed, single pole switch. Normally closed switches are drawn on a print in the closed postion (as shown in Figure 5.9).

The reason a switch is called normally closed is that in the rest position, the switch makes a connection, and power flows through it to the light. The light switch in your refrigerator is held in the open position by the closed door. No power flows through the switch while the door is closed, so the light stays off. When you open the door, the switch returns to its rest position (closed), and current can flow through the light.

The refrigerator light is an example of a normally closed switch that is open most of the time, since the door is usually closed. The terms "open" and "closed" (as opposed to "normally open" and "normally closed") refer to the state of the switch at any given instant. The PLC program can be set up to test whether any switch, whether normally open or normally closed, is open or closed at this exact instant, and to activate or deactivate an output based on the current postition of the switch.

## Photo Eyes and Proximity Detectors

Another thing you run into quite frequently on PLC motor logic is photo eyes and motion detectors. They are fairly prevalent on packaging machines and conveyor systems. As a matter of fact, both photo eyes and proximity detectors are used quite extensively.

Most times proximity detectors are set up to either detect a metal target or control a sorting gate, or that sort of thing. You can have a capacitive-type proximity unit that detects the product itself.

For product at a distance use a photo eye, of course. Photo eyes usually use a pulsating infrared-type signal, on the order of twenty or forty kilohertz, that won't feed a false trigger into your system whenever a regular (DC or a 60 Hz.) light shines on it.

Photo eyes and proximity detectors run into regular inputs like this:

**Figure 5.10.**

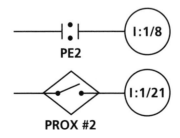

## Mechanically Interconnected Switches

Another thing you occasionally will run into, that you may not be familiar with, is the situation where you have a switch that is mechanically interconnected. Now on a regular electrician's diagram, a mechanically interconnected switch might look like this:

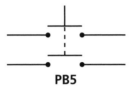

**PB5**

Figure 5.11.

On the print you will notice that the two elements of the switch are connected. When you put that switch into a PLC situation, a lot of the time you have a choice on how you want to wire that switch into your PLC. If, for example, the pushbutton has two poles (on this pushbutton they are mechanically interconnected), then there are two main choices.

One way you can program this switch on a PLC is to wire up both poles to an input terminal, similar to this:

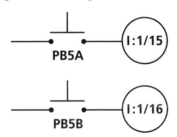

Figure 5.12.

The other way of doing it in your program is to use the same input for the switch and put it on two different rungs, similar to this:

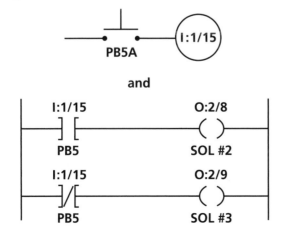

Figure 5.13.

What you ultimately end up with depends on how someone wires up the machine originally. Either way the machine is wired up, you can program the PLC to get the action that you want. That's the whole key to working with ladder logic.

What you are doing is taking the basic logic that electricians have used for years, and you are turning your AND, OR, NOT, etc. logic gate situations back into standard ladder logic. This is so that the people on the floor, who tend to be electricians, maintenance mechanics, and machine operators, can better interpret what's happening on their machines. These folks generally are the ones who have to do the troubleshooting, repair, or minor modifications to the machine. That's why ladder logic has become the mainstay in the PLC programming area, because you are trying to have the machines reflect an easy style of use for the people who are normally going to be working with them.

At this point in time, would you please turn to and do Exercise #5.

This exercise refers to timers, which aren't covered in detail until Chapter 7. The exercise itself is a good one for this chapter however. Just reference to a timer output coil for this exercise now and do the timer output rectangle later.

# Counters **6**

Many times, when dealing with machinery in manufacturing situations, you have to count items. Whether
you are putting a gross of pencils in a box or two dozen
ball bearings into a container, you need to count the
items coming off the machine.

## Count Up and Count Down Counters

Two major types of counters in the PLC are count
up counters (CTU) and count down counters (CTD).
In the general scheme of things, you probably will find
that count up counters work in 95% of all instances in
a PLC program. For the average person, count up
counters are much easier to work with than count
down counters. This is because you don't have to deal
with negative numbers and that sort of thing. For
most of what you want to do, you can use a count up
counter. So the next general rule is

## General Rule #6

Use count up counters, not count down counters.

## Preset, Reset, and Accumulated Values

When dealing with counters, you are going to have
to set them up for the PLC program. Basically, all PLCs
are set up the same way. Each counter of course has a
separate address. Each counter has a preset (a preset
value). Basically, the preset value is going to be how
many "things" you are going to count. Each counter
also has a reset. The reset typically is going to be zero,
because that is usually where the count starts from. It

doesn't have to start at zero, but for most people counting on their fingers, or whatever, they usually start from zero.

Your first count is going to be one (1). Hence, the reset is typically zero. The other item on a counter is the accumulator (the accumulated value). Typically this is also going to be zero. Usually that's how you want your accumulator to be in order to reflect the fact that you have nothing in it when you first start the counting. So you have a preset, a reset and an accumulator on a counter. Now on some PLCs some of these might be hidden off the screen (such as the reset value) or have default values, but that's fine.

Let's take a look at a typical circuit. Say we want to do something like count a baker's dozen ears of corn. We want to count them and then put them in a plastic bag. The bag is sized so that you want to count thirteen ears of corn to fill it.

Your first line would look like this:

**Figure 6.1.**

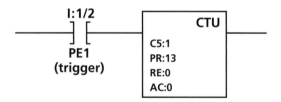

On the first line you have some sort of triggering mechanism, maybe from a photo eye or whatever. This sensor counts the ears of corn as they go by. You then go to the counter unit, where you set the preset at thirteen, where you set the reset at zero, and where you set the accumulated value at zero. This is all on the first line of the program.

Your second line would look like this:

**Figure 6.2.**

You would take a normally open contact from the counter and use it to toggle a solenoid.

If we look at the entire two line program now:

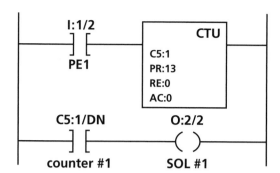

Figure 6.3.

We have the situation that the circuit is going to work by counting thirteen ears of corn. Then it's going to have the action occur of bagging the corn with the toggling of the solenoid. The problem with this two line program is that it will only do the action one time. The corn will keep coming down the line and the program will never re-bag. It just does its thing one time. It's sort of like what is called the "bomb maker circuit." That is to say, a circuit that only needs to work one time.

You have to add a third line, like this, for a reset.

Figure 6.4.

What will happen on this line is that, as the bag of product is finished up, it toggles another sensor of some sort. This action resets the accumulator. The counter goes back to zero and starts the process over again. You are going to count another thirteen ears of corn and bag them again.

## Enable and Done Bits

Basically there are two bits or flags that you use quite frequently with counters, and they are the Enable bit and the Done bit.

In the corn-bagging example, you notice that the timer contact is considered to have DN attached to it. DN is a Done bit. The Done bit comes on whenever you reach the value that is your preset (i.e., the accumulator value reaches the preset value).

The contact is also considered to have EN attached to it. EN is an Enable bit. The Enable bit comes on whenever there is a pulse going into your counter.

Now don't get me wrong; there are lots of other bits and flags that you can use in your counter. You have to look them up in your PLC manual for the particular aspects of how they might work. You have overflow bits and underflow bits and all kinds of bits, but by and large the Done and Enable bits are used in the vast majority of cases. They are the ones that you will be using.

## Example of a Light Chaser Circuit

The best way to illustrate how counters work is to do an actual circuit. The circuit we are going to do is what I call the light chaser circuit. Have you ever seen a movie marquee where the lights are flashing, and it looks like the lights are chasing each other around the movie marquee? Well, we are going to use a circuit with one pushbutton and five lights.

What the circuit is going to do is that, every time you push the pushbutton switch three times, a light is going to come on. Push it three more times and the second light is going to come on. Three more times and the third light comes on. This continues until you have all five lights on. Then when you push it three more times, all the lights go off and the sequence starts again. Three punches of the pushbutton switch and the

**Figure 6.5.**

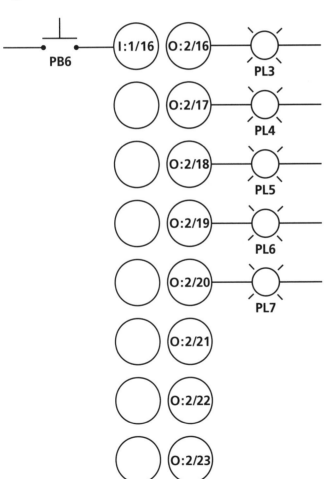

first light is on, three more punches and the second light comes on…This repeats over and over again.

So let's take a look at how that might be set up. One approach might be to have the pushbutton sensor (switch) be the trigger for the six counters. One counter would be used for each light, like this:

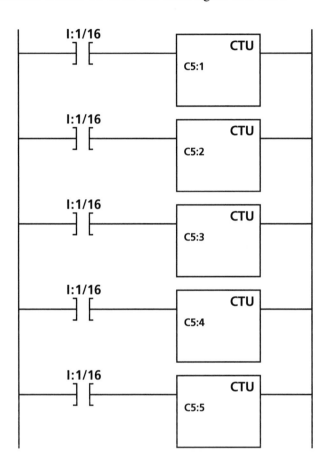

**Figure 6.6.**

What you would do then is to go through and set the presets on each counter. The first counter's preset would be three; the second counter would have a preset of six, then nine, twelve, fifteen, and eighteen. That way, each counter would have a trip point that would be three away from each of the other counters closest to it (Figure 6.7).

**Figure 6.7.**

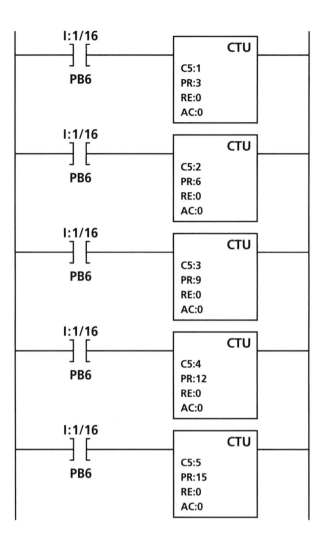

The next thing you would want to do is to add five more lines. Each line has a light and the counter driving that light, as in Figure 6.8.

Then add two more lines. When the sixth counter reaches its preset, it resets all six accumulators, so they go back to zero and turn all of the lights off, as in Figure 6.9.

Now go ahead and run this program and punch the pushbutton. You will notice that the sequence starts out OK and eventually turns on all five lights, and then starts all over again. However, in one of the program lines, if you actually count the number of pushes that you have to make for the light to go on through the se-

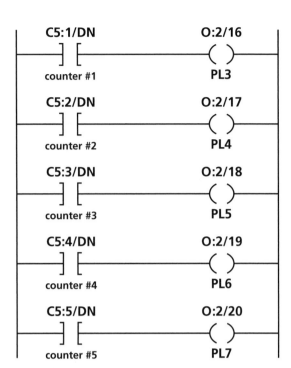

**Figure 6.8.**

quence, you will notice a slight problem: when you start over again at the beginning of the sequence, the count is off by one on the first light. This is due to a scan time problem you run into on PLCs.

When you are doing man-machine interfacing, the situation you have on the PLC is different from most hard wired electrical circuits. In a hard wired circuit, what happens is that basically the things in the circuit happen instantaneously. On the PLC, that appears to be the case, but what actually occurs is different. On the PLC, the whole program is scanned in about one sixtieth of a second. All your inputs are updated and then your outputs are updated, or your program is updated line by line, depending on how things are set up on your particular PLC. What's happening is that the PLC isn't instantaneously updated, it's updated line by line by line. You can have something occur say in line 17, and something else can occur in line 83, and within less then a sixtieth of a second you get a little glitch in your system. That glitch knocks the program off a little from what it actually needed to do, and this is the situation here with this program. You have a lit-

Figure 6.9.

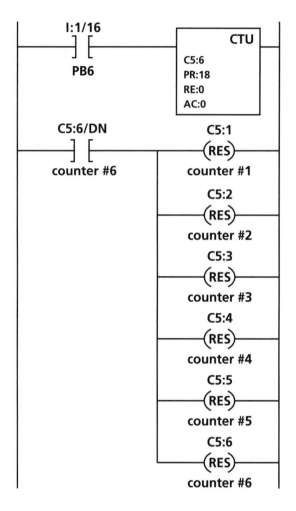

tle glitch that causes one of the counts to be off by one. As you are going through this circuit, your finger is going to be on the same button push both when you are resetting the counters and also when you are getting the first input for the first count on the first light. On most PLCs there is a one-shot timed contact. To take care of this problem, what you do is to insert the one-shot timed contact into the appropriate sensor line (usually the first counter program line), like this:

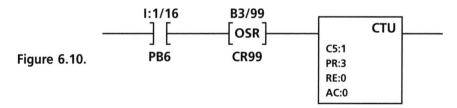

Figure 6.10.

That will typically take care of your problem. Now when we run this program, you will notice that the lights still come on in sequence OK, but now all the counts are proper.

Please turn to and do Exercise #6.

# Timers 7

When working in a manufacturing situation, the need arises to have various timing functions for various machine processes. The three main types of timers are on delay, off delay, and one-shot.

## Timer On Delay (TON)

The on delay timer is also known as delay on make or DOM. On the PLC, on delay is known as timer on delay (TON). It acts just like a burglar alarm system would. In a burglar alarm system you have a sensor on your window or your door, and you have an alarm circuit somewhere in your house. What happens if you break into a window or a door, or open a door, is that it breaks the circuit. Since it's an on delay system, the alarm is going to turn on eventually. You might have, say, thirty seconds or so to go over and disarm the system so that the alarm doesn't sound. It's called an on delay system because it delays going on.

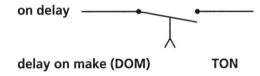

Figure 7.1.

## Timer Off Delay (TOF)

Another other type of timer is off delay, also known as delay on break or DOB. On a PLC it's known as timer off delay (TOF). This timer is like the action of the interior lights of a new car when you get out at night. When you turn off the engine at night, you get out of the car and the interior light stays on. The delay turns the lights off in about five or ten seconds, that sort of thing. What actually happens is that as soon as you toggle the switch on (by opening your door), you have an immediate action (your interior lights come on). But when you turn the switch off (by closing the door to your car), there is a delay (your lights delay going off). The delay is to allow for a certain amount of time to pass in order to give you enough time to see the path to your driveway or apartment door. It's called an off delay system or a delay on break (DOB).

**Figure 7.2.**

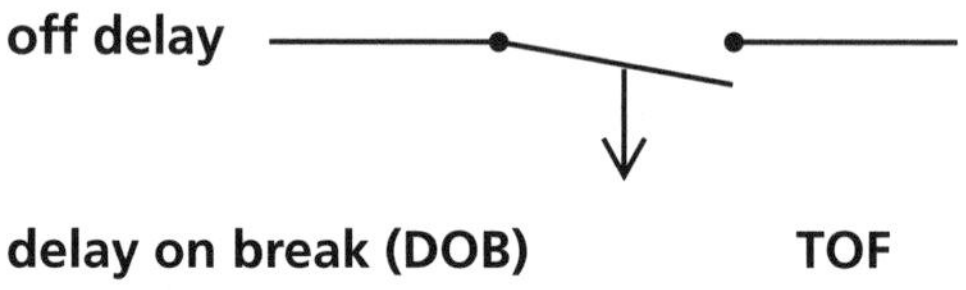

off delay

delay on break (DOB)                    TOF

## One-Shot Timer

The third type of timer is called a one-shot, but on the PLC there is not a specific timer called one-shot. You normally have to build a one-shot timer out of either an off delay or an on delay timer. What the one-shot is primarily used to do is to absorb or debounce any jitteriness in a circuit. It is used quite a bit as an interface between man and machine in many types of circuits you may run into. In computer systems, it is used very extensively where you have interfaces between the human-caused electro-mechanical inputs and the computer control processes.

one shot contact $\left[\text{OSR}\right]$

**Figure 7.3.**

one shot control relay CR3

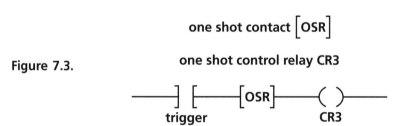

For instance, if you do not have a debounce circuit on a computer keyboard (which is basically a one-shot circuit), this is what would happen. If you go to the keyboard and, say, hit the letter A and smile knowingly, you are fooled. It appears that you hit the letter A key and that it clicks one time. But when you hit the letter A key, it does not actually make one click. It actually bounces several times. Well, more than several times. There are maybe 128 bounces to one key press.

If the computer responded to every bounce, what would happen when you hit the letter A would be that the whole screen would fill up with As. That would be crazy.

**Figure 7.4.**

So what they do is put a one-shot circuit either in hardware or software on the man-machine interfaces so that it ignores the additional bounces. On each key press during a certain time period, it just reads the multiple bounces of one key press as one bounce.

You have a similar situation when you have a roadway and there is that little sensor staked across the road to count the cars going by.

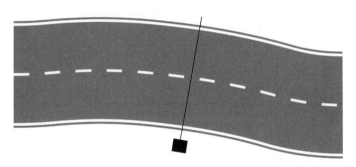

**Figure 7.5.**

The problem you have is that you want to count the vehicles, not the axles. You might have a car that comes by (blugh, blugh) that has two axles, and a truck that comes by (blugh, blugh, blugh,...blugh, blugh) that has multiple axles, and a kid in the neighborhood who comes by and just plays with the sensor to run the count up (blugh, blugh, blugh, blugh, blugh). You basically just want to count or read one vehicle regardless of the number of axles a vehicle has. No matter how many times the kids in the neighborhood are jumping up and down on the sensor, you want multiple inputs over a certain time period to count as one unit (vehicle). So you might pick a time period of a half-second, one second, or two seconds. Whatever you end up picking, no matter how many inputs you have at your sensor during that time period, you just read it as one output. That's called a one-shot.

## Preset, Reset, Accumulated Values, and Time Units

How are the timers put together on a PLC? They are a little different than a person would normally think of a timer. In most people's mind's-eye view, a timer is an analog type device that goes a certain length of time and then gives you an alarm, that sort of thing. Well, timers are not constructed that way on a PLC.

Timers are basically constructed just like counters. They are going to have a preset, a reset, and an accumulated value. The only basic difference between a timer and a counter on a PLC is that for the timers, instead of counting widgets (like you would be doing if you had a counter), you are going to count time units. The time unit on a particular counter might be one second, it might be a tenth of a second, or it might be one hundredth of a second.

Just for arguments sake, let's say the time unit on our particular timer on our particular PLC is a tenth of a second. If you wanted to time something for three seconds, what would your preset be? Well, if the time unit for this particular timer is a tenth of a second, that means your preset has to be thirty. You would have to count thirty time units with this timer to get three seconds. So the timer would look like this:

**Figure 7.6.**

(time base is .1 sec.)

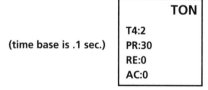

This timer would be one where your preset would be thirty, your reset would be zero, and the accumulator would be zero.

Now if I set the timer up in a circuit, like this,

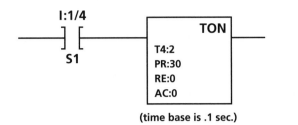

Figure 7.7.

the trigger would cause the timer to time out and toggle after three seconds.

Then in order to get an action out of the timer, I would have to add a second line that would look like this:

Figure 7.8.

When this timer timed out, the normally open contact from the timer's Done bit would toggle closed and cause the solenoid to operate. In this particular timer circuit (the on delay timer), if you removed the trigger pulse it would automatically reset the timer so that you can use it again. So for an on delay timer, you don't need to program a reset.

Of course you don't have to use an on delay timer. You can do a circuit as shown in Figure 7.9 which uses an off delay timer. You will notice that this particular circuit looks almost exactly like the on delay circuit. The difference in the off delay circuit is that when you

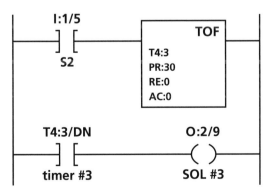

Figure 7.9.

trigger the timer, your output immediately takes place. Then three seconds after you remove the trigger, the output goes off. That's called the off delay timer. The reset for the off-delay timer is automatic, like the one in the on-delay timer.

If you need a one-shot timer, typically you have to build that out of an off delay timer or out of an on delay timer. If you actually want a full-fledged one-shot timer, you need to build it.

## One-Shot Timed Contact

Now there is a one-shot function on most PLCs, but usually it's what is called a one-shot timed contact. That's what we used in the last chapter. A timed contact lasts for just one scan cycle on the PLC. One shot timed contacts are just there to use when you have scan time problems that have created glitches in your system.

At this point in time, please turn to and do Exercise #7.

# Sequencers 8

In manufacturing processes, one of the things that you need to do is to have repeatability. You might have a box erector that keeps making the same type of box over and over again, or you might have a machine that makes a particular type of widget and it makes it over and over again. So a handy logic construction to have in a program is a sequencer. A sequencer is a logic device where you can take a set of inputs coming into a machine and go through a sequence of steps to a set of outputs, to make a box or a widget or whatever you want to do.

If you recall previously, we mentioned that a PLC was a technical device that was put together by committee. Well, sequencers were definitely put together by a committee. Sequencers are generally set up in a table format. You have things in the sequencer table from the electricians' world, from the regular people's world, and from the electronics and computer worlds. The net result is that you have a particular symbol on your PLC that allows you to do a repeating function over and over again with relative ease.

## Sequencer Symbol (SQO)

On the PLC screen, the sequencer circuit looks like a one line program, like this:

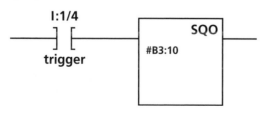

Figure 8.1.

SQO is the sequencer symbol and, in this case, #B3:10 is the sequencer's name. We will come back and add some other information to the symbol in a minute.

**Sequencers**

The sequencer can be either event-driven (by something passing in front of a photo eye or a switch, that sort of thing) or the sequencer can be time-driven (triggered or driven by a clock). It is very common for a sequencer to be driven either way. Just remember it can be either event or time-driven.

## Sequencer Output Table

What you need to do on most PLCs is to pull the sequencer output table out of the sequencer symbol block. By taking a look at the various parts in the sequencer table, you will be better able to understand how the sequencer works. Let's go ahead and look at the output table. It looks like this:

|  | ... | O:2/7 | O:2/6 | O:2/5 | O:2/4 | O:2/3 | O:2/2 | O:2/1 | O:2/0 |
|---|---|---|---|---|---|---|---|---|---|
| step 00 | | | | | | | | | |
| step 01 | | | | | | | | | |
| step 02 | | | | | | | | | |
| step 03 | | | | | | | | | |
| step 04 | | | | | | | | | |
| step 05 | | | | | | | | | |
| ⋮ | | | | | | | | | |

**Figure 8.2.**

You will notice in the figure that you have discrete outputs across the top row, starting with zero from right to left. You have steps in the first column going from top to bottom starting with double zero (00) and going however long you need to go. That's the basic gist of the table.

Now suppose your table, for instance, is on step number 01 and you have a one in two of the blocks, like this:

| | ... | O:2/7 | O:2/6 | O:2/5 | O:2/4 | O:2/3 | O:2/2 | O:2/1 | O:2/0 |
|---|---|---|---|---|---|---|---|---|---|
| step 00 | | | | | | | | | |
| step 01 | | 0 | 0 | 0 | 0 | 0 | 1 | 1 | 0 |
| step 02 | | | | | | | | | |
| step 03 | | | | | | | | | |
| step 04 | | | | | | | | | |
| step 05 | | | | | | | | | |
| ⋮ | | | | | | | | | |

**Figure 8.3.**

What that basically means is that when your sequencer goes to step 01, output one (O:2/1) and output two (O:2/2) are turned on because they have a binary one in that particular block where the row for step 01 and the columns for output one and output two intersect. There are zeros in all the other blocks on step 01. Your trigger then toggles, and the sequencer pulses and goes to step 02.

Step 02 might look like this:

**Figure 8.4.**

| | ... | O:2/7 | O:2/6 | O:2/5 | O:2/4 | O:2/3 | O:2/2 | O:2/1 | O:2/0 |
|---|---|---|---|---|---|---|---|---|---|
| step 00 | | | | | | | | | |
| step 01 | | | | | | | | | |
| step 02 | | 0 | 0 | 0 | 1 | 0 | 1 | 0 | 0 |
| step 03 | | | | | | | | | |
| step 04 | | | | | | | | | |
| step 05 | | | | | | | | | |
| ⋮ | | | | | | | | | |

You will notice on step 02 that output two is still on, output one has now turned off, and output four has come on. What generally happens is that the trigger will toggle the sequencer, and the sequencer will go down through this table step by step by step. The table then toggles signals to the particular outputs your row happens to turn on. If you have a one at an intersecting column and row in that particular step, then it turns the output on. If you have a zero in the particular step's row and output column, then the zero will turn that output off.

A regular table on a typical PLC would look something like this:

**Figure 8.5.**

| | ... | O:2/7 | O:2/6 | O:2/5 | O:2/4 | O:2/3 | O:2/2 | O:2/1 | O:2/0 |
|---|---|---|---|---|---|---|---|---|---|
| step 00 | | 0 | 0 | 0 | 0 | 0 | 0 | 0 | 0 |
| step 01 | | 1 | 0 | 0 | 0 | 0 | 1 | 1 | 0 |
| step 02 | | 1 | 0 | 0 | 1 | 0 | 1 | 0 | 0 |
| step 03 | | 0 | 1 | 0 | 1 | 0 | 1 | 0 | 0 |
| step 04 | | 0 | 1 | 0 | 1 | 0 | 0 | 1 | 0 |
| step 05 | | 1 | 0 | 0 | 1 | 0 | 0 | 1 | 0 |
| ⋮ | | | | | | | | | |

## File, Mask, Destination, Control, Length, and Position Values

An actual circuit might look like this:

**Figure 8.6.**

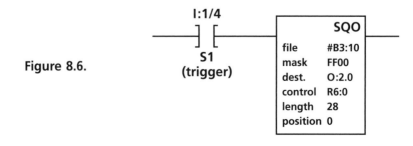

The file is the name of the sequencer. For our purposes, we are going to call the sequencer #B3:10. Parts of the name in most PLCs mean something specific. They are like an address in the memory map.

The mask lets you describe what terminals you want that sequencer to use. You may have a card on the output rack that has 32 spots (terminals) on it. Well, your sequencer may not necessarily need the whole 32 spots. You may have used 8 of them so far (in the other parts of the program that have nothing to do with the sequencer). What you can do with the mask is to mask out the first 8 spots that you have already hooked up wires to, so that you wouldn't use them in the sequencer table. So if you had a 32 bit sequencer and you have already used the first 8 terminals which are 0–7, you would then use terminals 8–31 in your sequencer. Your mask would be FFFFFF00, which of course is in hexadecimal (h) format. What that mask will do for you is block out the first 8 bits so the table won't effect them, but lets you use the other 24 bits which would correspond to terminals 8–31. Your mask would be on the table in binary and it would look like this:

**Figure 8.7.**

| Output | ... | 16 | 15 | 14 | 13 | 12 | 11 | 10 | 9 | 8 | 7 | 6 | 5 | 4 | 3 | 2 | 1 | 0 |
|---|---|---|---|---|---|---|---|---|---|---|---|---|---|---|---|---|---|---|
| Mask | | 1 | 1 | 1 | 1 | 1 | 1 | 1 | 1 | 1 | 0 | 0 | 0 | 0 | 0 | 0 | 0 | 0 |
| Step 00 | | | | | | | | | | | | | | | | | | |
| Step 01 | | | | | | | | | | | | | | | | | | |
| Step 02 | | | | | | | | | | | | | | | | | | |
| Step 03 | | | | | | | | | | | | | | | | | | |
| Step 04 | | | | | | | | | | | | | | | | | | |
| Step 05 | | | | | | | | | | | | | | | | | | |
| ⋮ | | | | | | | | | | | | | | | | | | |

The next thing up on the screen is the destination, and in our particular case we are going to go to slot two to the output card we have there (O:2.0). This means output the sequencer to the output card in slot #2 of the PLC.

The next thing listed is control. That would be the memory location in the control file where your control bits and flags are located (such as the Done bit, etc.).

The length is basically how many steps you want to have in your sequencer table. For example, your table might need to be 20 steps long in order to do the actions you want to do.

The position that you are going to start your table from is almost always position zero (Step 00). Depending on the PLC, you might also need to specify a home row position. You just have to see how your particular PLC works or look up sequencers in the PLC's manual.

## Example of a Light Chasing Circuit

Now the best way that I know of to see how sequencers operate is to do one. That way you can see how the results work together. The one we are going to do is the same circuit that we did back in the counting section. The one where the marquee lights are following themselves around. The one where you push the button three times and it lights one light, push it another three times and the second light comes on, etc. We are going to put that circuit action in the sequencer and initially make it event-driven. Then we are going to take the same sequencer circuit and change the trigger and make the sequencer time-driven.

Let's go ahead and set up the screen. The screen would look something like this:

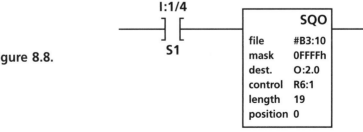

Figure 8.8.

Your table would end up looking something like this:

| output | ... | O:2/7 | O:2/6 | O:2/5 | O:2/4 | O:2/3 | O:2/2 | O:2/1 | O:2/0 |
|---|---|---|---|---|---|---|---|---|---|
| mask | | 1 | 1 | 1 | 1 | 1 | 1 | 1 | 1 |
| step | | | | | | | | | |
| 00 | | | | | | | | 0 | |
| 01 | | | | | | | | 0 | |
| 02 | | | | | | | | 0 | |
| 03 | | | | | | | | 0 | |
| 04 | | | | | | | | 1 | |
| 05 | | | | | | | | 1 | |
| 06 | | | | | | | | 1 | |
| 07 | | | | | | | 1 | 1 | |
| 08 | | | | | | | 1 | 1 | |
| 09 | | | | | | | 1 | 1 | |
| 10 | | | | | | 1 | 1 | 1 | |
| 11 | | | | | | 1 | 1 | 1 | |
| 12 | | | | | | 1 | 1 | 1 | |
| 13 | | | | | 1 | 1 | 1 | 1 | |
| ... | | | | | | | | | |

| output | ... | O:2/7 | O:2/6 | O:2/5 | O:2/4 | O:2/3 | O:2/2 | O:2/1 | O:2/0 |
|---|---|---|---|---|---|---|---|---|---|
| mask | | 1 | 1 | 1 | 1 | 1 | 1 | 1 | 1 |
| step | | | | | | | | | |
| 00 | | 0 | 0 | 0 | 0 | 0 | 0 | 0 | 0 |
| 01 | | 0 | 0 | 0 | 0 | 0 | 0 | 0 | 0 |
| 02 | | 0 | 0 | 0 | 0 | 0 | 0 | 0 | 0 |
| 03 | | 0 | 0 | 0 | 0 | 0 | 0 | 0 | 0 |
| 04 | | 0 | 0 | 0 | 0 | 0 | 0 | 1 | 0 |
| 05 | | 0 | 0 | 0 | 0 | 0 | 0 | 1 | 0 |
| 06 | | 0 | 0 | 0 | 0 | 0 | 0 | 1 | 0 |
| 07 | | 0 | 0 | 0 | 0 | 0 | 1 | 1 | 0 |
| 08 | | 0 | 0 | 0 | 0 | 0 | 1 | 1 | 0 |
| 09 | | 0 | 0 | 0 | 0 | 0 | 1 | 1 | 0 |
| 10 | | 0 | 0 | 0 | 0 | 1 | 1 | 1 | 0 |
| 11 | | 0 | 0 | 0 | 0 | 1 | 1 | 1 | 0 |
| 12 | | 0 | 0 | 0 | 0 | 1 | 1 | 1 | 0 |
| 13 | | 0 | 0 | 0 | 1 | 1 | 1 | 1 | 0 |
| ... | | | | | | | | | |

Go ahead and write this program and put it into your PLC. Of course you need to remember that the actual number of steps in your sequencer and how you do the mask might vary on your PLC, but the table will generally look almost exactly the same. Go ahead and put this program into your PLC and run it.

You will notice that when you press the button, the program is going to go through all the steps one at a time. On the PLC output card the first light comes on, and stays on for three pulses. The second light comes on during the next three pulses (light one and two are both on). This continues all the way to the end of the table where all five lights are on.

Now one of the characteristics of a sequencer table is that it is going to fall through the bottom of the table if you keep putting pulses in. The sequence is going to fall through the bottom of the table and automatically start over at the top again. The whole object of having the sequencer table is that it works as a repeating type function. Whether you are making boxes in a box erector or making widgets in a machine, you are doing the same action over and over and over again.

Now we are going to do the same thing but make the sequencer time-driven. You don't have to change the table at all.

All you have to do is set up a clock to drive the sequencer on the screen in a way that your program will get a clock to pulse every three seconds or whatever. The program line for the clock would look like something similar to this:

**Figure 8.10.**

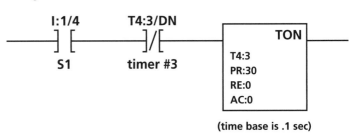

(time base is .1 sec)

Then in the program line that includes the sequencer, instead of having the pushbutton switch as a trigger, you simply replace the pushbutton switch with the Done bit normally-open contact from the timer, like this:

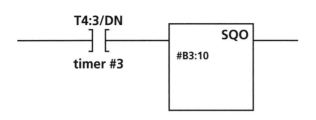

**Figure 8.11.**

You would end up with a two line program that looks like this:

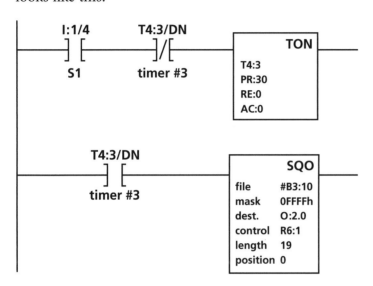

**Figure 8.12.**

When you close the switch (the trigger for the timer), the timer is going to run. As the program scans through these two lines, every three seconds the timer is going to pulse and reset itself. When it pulses and resets itself, it's also going to pulse the sequencer and the sequencer is automatically going to step through its table. The lights are going to come on every three seconds. The sequencer is going to keep rotating around through the table doing the light chasing thing over and over again until you remove the trigger.

Would you please turn to and do Exercise #8.

# Math Functions **9** and Analog Inputs and Outputs

Besides the discrete inputs and outputs you have with the PLC, you also have analog inputs and outputs. Math functions are particularly useful when working with analog inputs and outputs.

## Addition, Subtraction, Multiplication, and Division Functions

The math functions are fairly straight forward. Typically, the most used functions are the ones off of the four banger calculator, where you add, subtract, multiply and divide.

So for example, you can take the output from two accumulators and add them together like this:

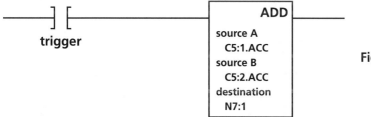

Figure 9.1.

In a similar fashion you could use subtraction, multiplication, and/or division. The easiest way, of course, is to use whole numbers (integers), which most of the time are all that you need.

## Less Than, Greater Than, and Equal To Functions

In addition to the four banger calculator features that are available, three equalities are also very handy. These equalities are less than, greater than, and equal to.

A good example for using the equalities greater than and less than would be to do something similar to a deadband circuit on an air conditioning unit. There you basically have a situation that if the temperature gets over 72 degrees F, you want the compressor of the air conditioning to start up and the unit to come on. You do not want the A/C unit to shut off until it gets down to 68 degrees F or below. Between 68 and 72 degrees F, you do not want any change of state going on. This no-change zone gives you a deadband in the circuit. (The official name for deadband is hysteresis.) The deadband is in order to prevent the compressor from cycling all the time (going on and off). You don't want to have the compressor cycling all the time because if it does do this, it is drawing 6–8 times its normal operating current because the turning on's draw a lot of current when they first get the compressor going. Cycling all the time would end up overheating the compressor.

## Analog Inputs and A-to-D Conversion

You have available on most PLCs various analog cards to give you analog inputs and outputs. You have various voltage and current ranges available (typically 0–5 volts input and output, 0–10 volts input and output, or 4–20 milliamps current input and output). The biggest thing about using analog is conversion. If you're hooking up the wires to a sensor and you have an analog signal coming into the PLC, it goes through an analog-to-digital (A-to-D) converter on a PLC input card that calculates some number that corresponds to the corresponding signal.

So for instance, you might have a 0–10 volt circuit, and 7 volts coming in off of a particular thermostat might be equivalent to 100 degrees Fahrenheit. In the

PLC that might be equal to the number 8000, because there is a correlation between how many volts are coming in and what number corresponds to it in the PLC. You can look on a chart for the particular PLC you are operating and find out what the correspondence is between input voltage and the number. Once you have a number in the PLC, you can use math functions to manipulate it and you can create input or output signals with it.

## Analog Outputs and D-to-A Conversion

You have a similar thing going on when you output an analog signal. You might have a situation where you have a variable frequency drive looking for a 0–10 volt signal that corresponds to different speeds on an AC motor. You may have a situation that at full speed your motor is running at 1740 RPM, and you want to run it at about half speed or at 800 RPM. This corresponds to maybe about 4000 in your PLC. Your PLC takes the number in the PLC and does a digital-to-analog (D-to-A) conversion of the number into a voltage. The voltage might be approximately 4.5 volts, and it goes out to the variable frequency drive through the PLC's analog card. The variable frequency drive takes that voltage signal and translates it through its circuitry. It then outputs (to the motor) a frequency of 30 Hz, instead of 60 Hz. The motor will now run at about 800 RPM.

## Example of a Deadband Circuit

The best way to see how all of this works together is to do a circuit. We are going to go ahead and do the aforementioned deadband circuit on the air conditioning unit. We are going to have a fairly straightforward circuit here. We are going to have the temperate sensor come into an analog to digital card and we are going to calibrate the sensor at two points. One point is freezing water at 32 degrees Fahrenheit, and the second is boiling water at 212 degrees Fahrenheit.

Then what we want to do in the PLC is have the sensor set up so that when it senses 72 degrees Fahrenheit, the PLC is going to turn on the compressor. Once the sensor goes to 68 degrees Fahrenheit or below, the PLC is going to turn off the compressor, and the PLC will not turn the compressor back on until the temperature gets above 72 degrees Fahrenheit.

**Figure 9.2.**

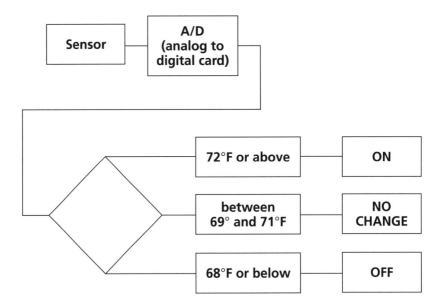

This is your typical deadband type circuit.

In order to do the circuit you are going to need a temperature sensor, one analog to digital card, a PLC, and some sort of output like a contactor coil. For the output that is going to turn the compressor on and off, you can use a pilot light or some type of output contactor coil. Of course you will also have a one-time need of some water with ice for one of the calibration points, and boiling water (maybe using a coffee maker or what not) for the other point of calibration.

In this particular program, it is very easy to use the equality statements (be it less than, greater than, or equal to) to run the circuit. A typical circuit to turn on the compressor when the temperature gets over 72 degrees would use greater than and look like this:

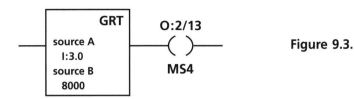

Figure 9.3.

And a typical circuit to turn the compressor off if it went below 68 degrees would use less than and would look like this:

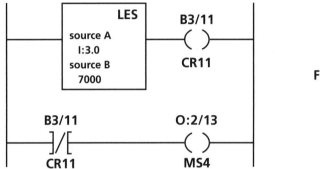

Figure 9.4.

Of course to run the actual circuit you would have to combine these two circuit rungs, and you could do this by using a control relay type circuit like this:

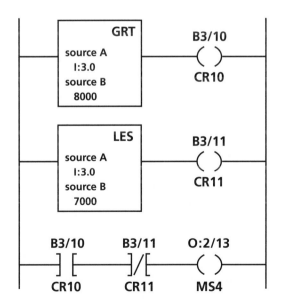

Figure 9.5.

Then use a latch and unlatch coil to add the dead-band, like this:

**Figure 9.6.**

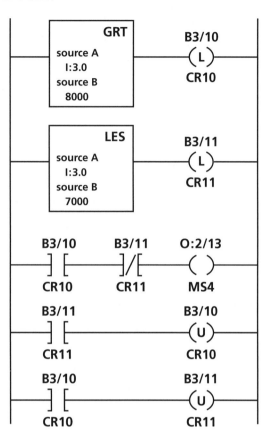

In summary, the basic key to doing these types of circuits is to stay with your basic math functions, check your calibration points, and follow what I call general rule #7. This rule has existed in engineering for a long long time and it's the kiss rule.

## General Rule #7

**Keep It Simple, Stupid.**
Where most people get messed up when dealing with this type of circuitry is that they try to get too complicated. They end up by not controlling the situation. They let the situation control them. They do not know what to control in the fashion they want it to be controlled. I have found that the best way to go about

it in control circuits is to: stay with your standard math functions, keep accurate calibration points, and make sure your sensors are shielded if necessary and hooked up properly. Typically you are not going to have any problems that way.

At this point in time, please turn to and do Exercise #9.

# Appendix A

**Symbols in the Electrical World**

*Switches*

**Figure 1.**

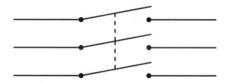

**DPST**
two position switch
(double pole
single throw)

**3PDT**
two position switch
(three pole
double throw)

**3PST**
two position switch
(three pole
single throw)

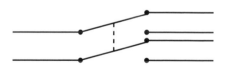

**DPDT**
two position switch
(double pole
double throw)

**SPDT**
two position switch
(single pole
double throw)

**Figure 2.**

**SPST**
**two position switch**
**(single pole single throw)**

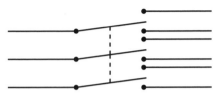

**3PDT**
**three position center off switch**
**(three pole double throw)**

**normally open**
**momentary contact**
**single pole**
**pushbutton switch**

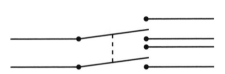

**DPDT**
**three position center off switch**
**(double pole double throw)**

**SPST**
**normally closed switch**
**(two position)**

*Contacts and Switches with Flags*

**Figure 3.**

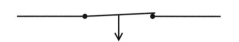

**normally closed**
**off delay**
**timed contact**

**normally open**
**on delay**
**timed contact**

**normally closed**
**on delay**
**timed contact**

**normally open**
**off delay**
**timed contact**

**Figure 4.**

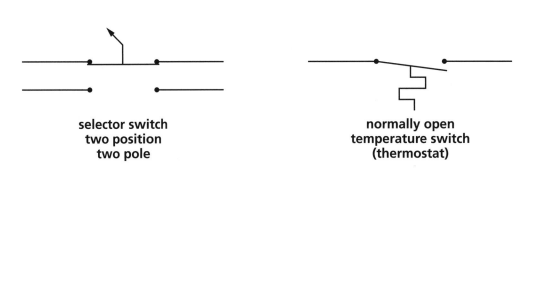

**selector switch
two position
two pole**

**normally open
temperature switch
(thermostat)**

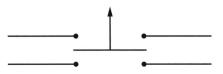

**normally closed
pressure (or vacuum)
switch**

**selector switch
three position
two pole**

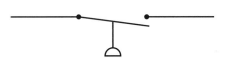

**normally open
pressure (or vacuum)
switch**

Figure 5.

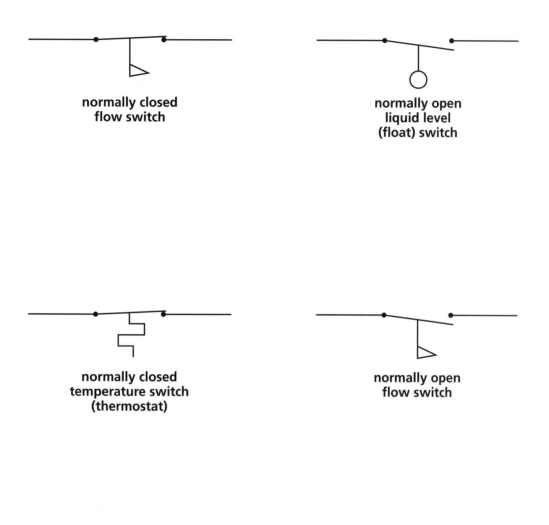

normally closed
flow switch

normally open
liquid level
(float) switch

normally closed
temperature switch
(thermostat)

normally open
flow switch

normally closed
liquid level
(float) switch

**Figure 6.**

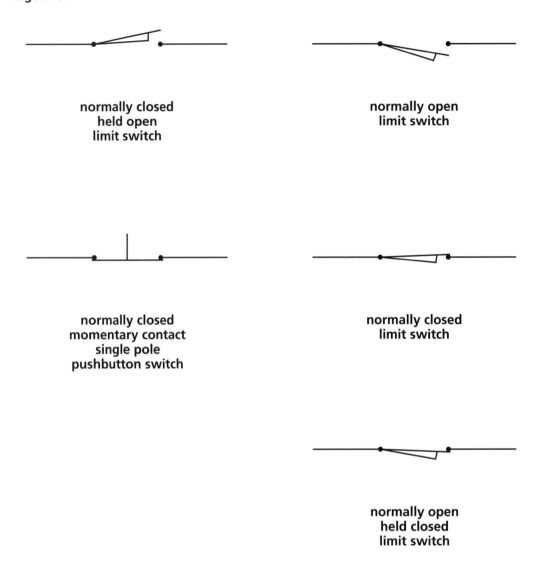

normally closed
held open
limit switch

normally open
limit switch

normally closed
momentary contact
single pole
pushbutton switch

normally closed
limit switch

normally open
held closed
limit switch

**Figure 7.**

normally closed contact

normally open contact

*Coils and Loads*

**Figure 8.**

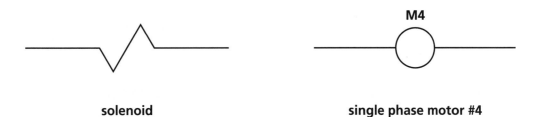

solenoid
(series coil)

**M4**

single phase motor #4

**CR3**

control relay coil #3

**MS2**

motor starter coil #2

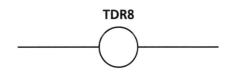

**TDR8**

time delay relay #8

**Figure 9.**

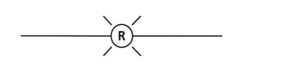

**R11**

**resistor #11 (schematic)**

**resistor #13 (pictorial)**

**red pilot light
(or red indicator light)**

**3 phase motor (3ø motor)**

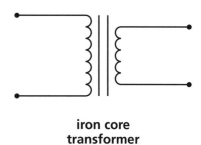

**iron core
transformer**

*Overcurrent Protection and Disconnects*

**Figure 10.**

**2 pole, single phase, (1ø)
circuit breaker**

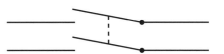

**2 pole, single phase, (1ø)
disconnect**

O/L's

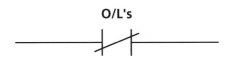

**normally closed
overload relay contacts**

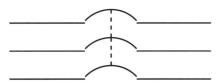

**3 pole, 3 phase, (3ø)
circuit breaker**

**single pole, single phase, (1ø)
circuit breaker**

**Figure 11.**

fuse (schematic)                    thermal overload
                                         (or heater)

fuse (pictorial)                    3 pole (3ø)
                                         disconnect

*Miscellaneous*

**Figure 12.**

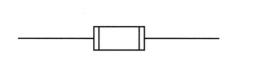

diode (rectifier)

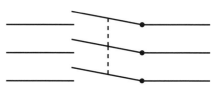

L1      L2      L3

incoming power
(line side)

earth ground

T1      T2      T3

outgoing power
(load side)

**Figure 13.**

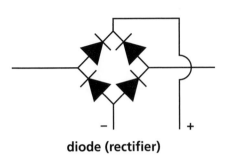

**diode (rectifier)**

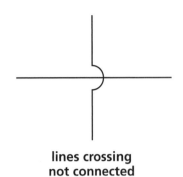

**lines crossing
not connected**

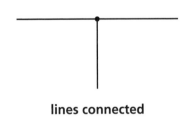

**lines connected**

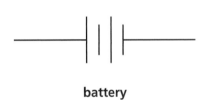

**battery**

- - - - - - - - - - - - - - - - - - - - -

**mechanically
interconnected**

# Symbols in the Electronic World

*Solid State Diode*

Figure 14.　anode  cathode

*SCR*

Figure 15.　anode 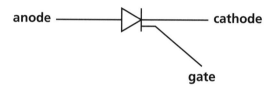 cathode

gate

*Triac*

Figure 16.　MT2 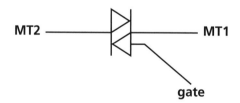 MT1

gate

*LED*

Figure 17.

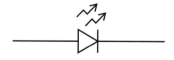

*NPN Transistor*

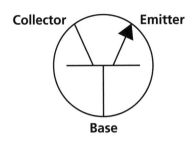

Figure 18.

*PNP Transistor*

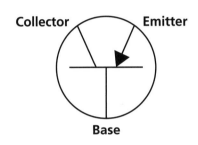

Figure 19.

*Regular Resistor*

Figure 20.

*Pot*

Figure 21.

*Photo-Eye*

Figure 22.

*Chassis Ground*

**Figure 23.**

*Earth or Chassis Ground*

**Figure 24.**

## Logic

*AND*

**Figure 25.**

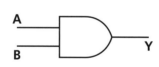

| A | B | Y |
|---|---|---|
| 0 | 0 | 0 |
| 0 | 1 | 0 |
| 1 | 0 | 0 |
| 1 | 1 | 1 |

*OR*

**Figure 26.**

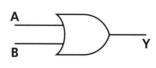

| A | B | Y |
|---|---|---|
| 0 | 0 | 0 |
| 0 | 1 | 1 |
| 1 | 0 | 1 |
| 1 | 1 | 1 |

*NOT*

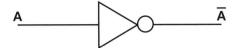

Figure 27.

*YES*

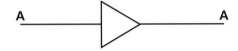

Figure 28.

*NAND*

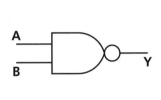

| A | B | Y |
|---|---|---|
| 0 | 0 | 1 |
| 0 | 1 | 1 |
| 1 | 0 | 1 |
| 1 | 1 | 0 |

Figure 29.

*NOR*

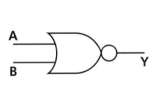

| A | B | Y |
|---|---|---|
| 0 | 0 | 1 |
| 0 | 1 | 0 |
| 1 | 0 | 0 |
| 1 | 1 | 0 |

Figure 30.

# Appendix B

This appendix contains computer logic diagram equivalents for the figures in chapter 4.

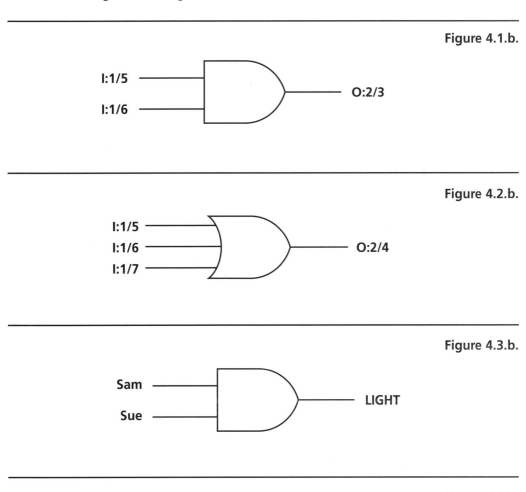

**Figure 4.1.b.**

I:1/5 ──┐
        ├── O:2/3
I:1/6 ──┘

**Figure 4.2.b.**

I:1/5 ──┐
I:1/6 ──┤── O:2/4
I:1/7 ──┘

**Figure 4.3.b.**

Sam ──┐
      ├── LIGHT
Sue ──┘

**Figure 4.4.b.**

Sam ────┐
Sue ────┤── HORN
George ─┘

**Figure 4.5.b.**

**Figure 4.7.b.**

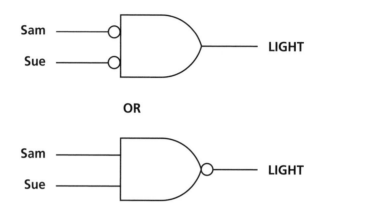

OR

**Figure 4.8.b.**

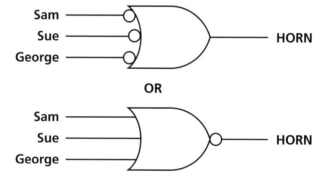

**Figure 4.9.b.**

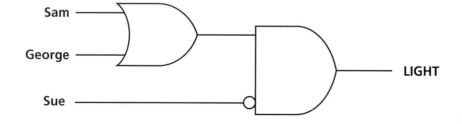

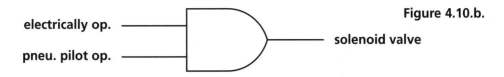

Figure 4.10.b.

electrically op. ———

pneu. pilot op. ———

——— solenoid valve

Figure 4.11.b.

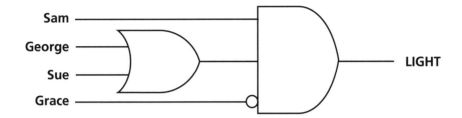

Figure 4.12.b.

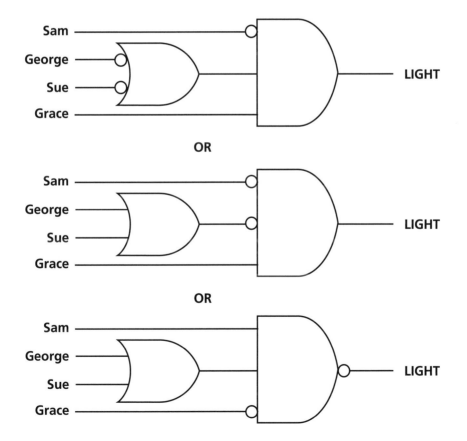

**Figure 4.13.b.**

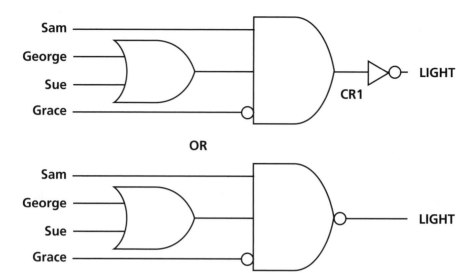

OR

# Exercises

Put the following circuits or situations into a PLC program by 1) assigning a terminal number to each device, and 2) in the ladder logic, replace each symbol and name with the equivalent PLC symbol. The result is your first PLC programs:

## Exercise #1

Figure 1.

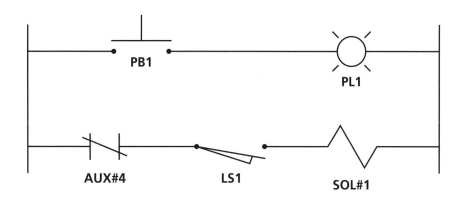

## Exercise #2

Figure 2.

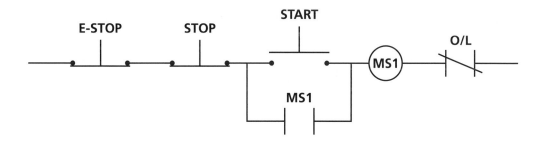

## Exercise #3

**Figure 3.**

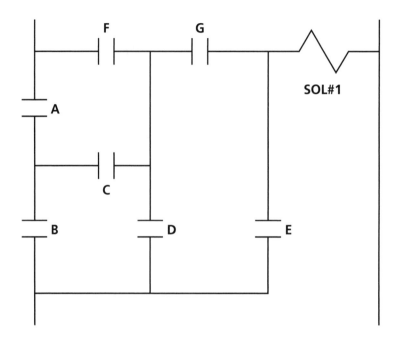

## Exercise #4

If Grace and Sam but not Sue, or George comes in, then open up the till.

## Exercise #5

An oiler runs as soon as a compressor starts up and for 90 seconds afterward. If the temperature is too high the oiler comes on, regardless of what the timer says.

## Exercise #6

You have a parking garage with two entrances and two exits. Your garage has a capacity of 200 cars. When you reach 180 cars a warning light comes on in the attendants booth. When you reach 195 cars a sign comes on that says "Lot Full".

## Exercise #7

You have four solenoids numbered 1, 2, 3, and 4. When you activate your machine solenoids #1 and 3 turn off. Ten seconds later solenoids #2 and 4 turn on. If you shut the machine down, solenoids 2 & 4 turn off immediately, while solenoids 1 & 3 delay turning back on for eight seconds.

## Exercise #8

Make a sequencer to mimic your washing machine cycle. Use a step time of one second instead of one minute. Use the following outputs:

- hot water valve
- cold water valve
- drain valve
- wash speed
- spin speed

## Exercise #9

Get a thermistor from a building products or appliance store and calibrate it to a PLC analog/digital card.

# Selected Solutions

Did you come up with something like this?

**Exercise #1**

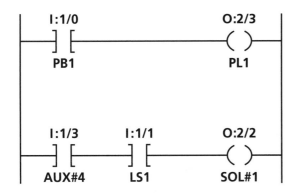

**Figure 1.**

**Exercise #2**

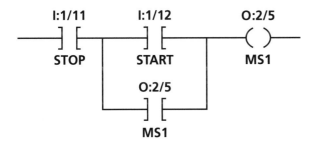

**Figure 2.a.**
**PLC Screen**

**Figure 2.b.**
**PLC terminal diagram.**

**Figure 2.c.**

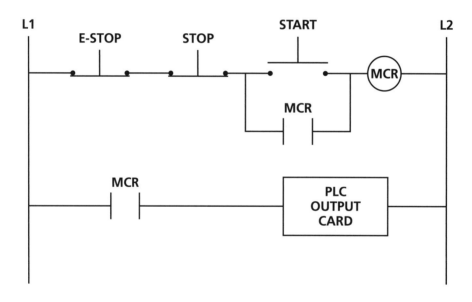

## Exercise #3

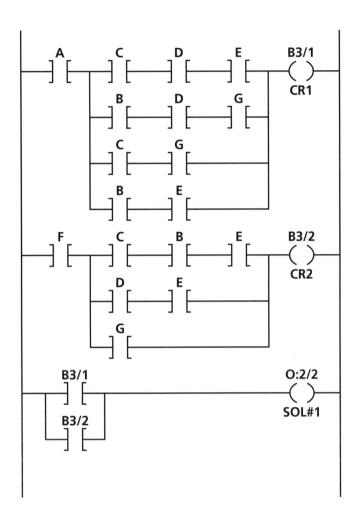

Figure 3.

## Exercise #4

Figure 4.

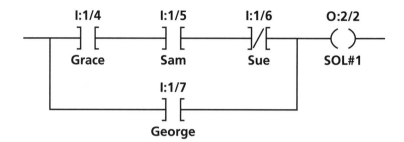

## Exercise #5

Figure 5.

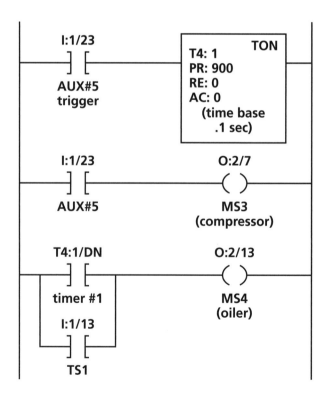

**Notes**